光のゼミナール

THE LIGHT SEMINAR

光のゼミナール

THE LIGHT SEMINAR

武蔵野美術大学 空間演出デザイン学科 面出薫ゼミ 10年間の記録

Ten years activities of Kaoru Mende Seminar
Department of Space Design, Musashino Art University

鹿島出版会

面出薫+ゼミ編集委員会 編著

光のゼミナール
目次

	8	序論: 光のゼミナール 10年間の試行 面出薫	光を操る	155
			158	新潟 岩室温泉 ― 光のまちづくり
			164	新正卓展照明計画
	16	光のデザイン ― 7つの学び方	168	江戸東京たてもの園 ― ライトアップ
			184	歌舞伎町ルネッサンス
光を観る	19		190	個人課題 ― 卒業制作
	22	世界の光を観る		
	26	照明探偵団とは何だ?	座談会	205 座談会1: 卒業生たち
	28	光の世界旅行 ― ニューヨーク 2005		卒業生がゼミで学んだこと
	30	光の世界旅行 ― シンガポール 2006		221 座談会2: 面出教授＋講師・助手
	32	光の世界旅行 ― ベオグラード 2008		ゼミで伝えたかったこと

光を集める 39
　　　　　42　ライトコレクション
　　　　　54　光に詩を添える

236　あとがき

238　138人の面出ゼミ卒業生

光に触れる 61
　　　　　64　ライトアップゲリラ
　　　　　76　ライトボックスの制作

コラム
　　　　　58　杉本貴志
　　　　　　　面出ゼミの誕生とその意味

自然光に学ぶ 83
　　　　　86　伝統民家の光 ― 五箇山合宿
　　　　　94　緑と海の合宿 ― 清里合宿・大島合宿
　　　　　102　オーロラ観測隊

122　原 研哉
　　　面出薫 火と光の記憶

闇を知る 107
　　　　　110　明治神宮の闇に学ぶ
　　　　　112　夜の絵はがきづくり
　　　　　118　闇のライトアップ

152　佐藤 卓
　　　光で世の中を感じる

188　小竹信節
　　　完璧をつくる隙間

影と遊ぶ 125
　　　　　128　キャンドルナイト ― 東京・表参道
　　　　　146　キャンドルナイト ― シンガポール

202　小池一子
　　　先生は現場を用意する

THE LIGHT SEMINAR

Contents

Preface   8   The light seminar ten years of experiments
                      Kaoru Mende

                 16   Seven ways to study the design of light

Observing Light   19

                 22   Observing light all over the world
                 26   What is Lighting Detctives?
                 28   World lighting tour in New York 2005
                 30   World lighting tour in Singapore 2006
                 32   World lighting tour in Belgrade 2008

Collecting Light 39

                 42   Light collection
                 54   Writing a poem for light

Touching Light   61

                 64   Light-up guerrilla
                 76   Making of light box

Learning from Natural Light   83

                 86   Light in the traditional private house
                 94   Annual camps with green and sea in Kiyosatoand Ohshima
                 102 Aurora observation team

Knowing Darkness   107

                 110 Learning from darkness in Meiji Shirine
                 112 Making "Night picture postcards"
                 120 Light-up the darkness

Playing with Shadows   125

                 128 Candle Night in Omotesando, Tokyo
                 146 Candle Night in Singapore

Manipulating Light   155

                 158 Iwamuro hot spring town, Niigata — Town planning by light
                 164 Light planning for the exhibition by Taku Aramasa
                 168 Edo Tokyo Open Air Architectural Museum — Light-up
                 184 Kabukicho renaissance
                 190 Assignments for individuals — Graduation Work

Discussion meeting   205 Discussion meeting1:
                            What alumni learned from Mende seminar

                 221 Discussion meeting 2:
                      What we tried conveying to students

                 236 Afterword

                 238 138 alumni of Mende seminar

Column

                 58   Takashi Sugimoto
                      The Birth ans Significance of Mende Seminor

                 122 Kenya Hara
                      Kaoru Mende — Memories of Fire and Light

                 152 Taku Satoh
                      Sening the World though Light

                 188 Nobutaka Kotake
                      The Gap That Creates Perfection

                 202 Kazuko Koike
                      Preparation of the Site by Teachers

序論:光のゼミナール ── 10年間の試行

面出 薫

光のデザインという道にはまってから35年を迎えようとしている。1978年に照明デザイナーになろうと決心した時、皮肉なことに私は照明デザインという職能に対して何らかのうさん臭さを感じていた。その道の先達には大変失礼な話だが、それは照明デザインが豪華でキラキラしたものの代名詞であり、店先に並べられる光の家電商品そのものを意味していたからに違いない。プロダクトデザインを大学で学んでいながら、それが詰まる所、産業廃棄物というゴミの山を作る行為のように思えて、私はそこから逃れたいと考えていた矢先だった。

だから私は、今さら「どうして照明デザイナーになったのですか」と聞かれることが苦手で仕方ない。立派な理由もなく、あえて正直に告白するなら「私が目指したいと思った環境デザインの仕事が見つからなかったから、仕方なしに…」と告白する。これはデザイナーにとってあまりに美談ではない。しかし、皮肉なことに批判的であったはずの照明デザインの世界に入り、それを「光のデザイン」といい換えてみると気持ちがすっきりした。光環境のデザインという視点に気づいた時に光明を見た感じがして、照明器具の姿かたちでなく光や闇と人間との深遠な関係をデザインできるとしたら、それは無限の世界への挑戦でしかない。楽しく険しい光のデザインの未来を同時に予感した。

どのように光のデザインを学んだか

それまでに光のデザインという分野の勉強など微塵もやってこなかったので、私は照明会社の研究所に入社して初めて光のデザインを学び始めた。既にアメリカ東部で成立していた建築照明デザインというジャンルを知ってからはとりわけ、世界中の自然光の現場に赴き、世界中の建築空間と光を体感することに努力した。右肩上がりの高度経済成長期がそれをバックアップしてくれて、照明会社の研究所は多少英語が堪能だった私を、世界各国に派遣し調査する仕事を与えてくれた。建築の本や雑誌を夢中に読み漁り、世界の照明カタログを分析し、照明学会編の技術書を何回も反復して学習したが、文字に書かれているセオリーは無防備に信用しなかった。理屈上の解説は自分で光の効果を体験した時にのみ納得し、時に否定もした。私は正直な体験主義者だった。新たなものを見れば見るほど、自分の眼や脳や体内に様々な光の様相が染み込んでいく。その感覚を大切にした。アメリカ東部で盛んだった照明コンサルタントとの仕事を共働する機会に恵まれ、彼らの良い仕草や技だけを真似していった。E.サーリネン、ミース・F.D.ローエ、ルイス・I.カーン、P.ジョンソン、I.M.ペイなど、アメリカの近代を作った建築照明の仕事には何度となく足を運び心を打たれた。そして近代以前の歴史的建築、日本の数寄屋建築や神社仏閣に宿る光でさえ、それぞれの文化が育んだ光環境の所作であることを知る。目にするものすべてが光のデザインのための学習過程であったのだ。

どうして光のゼミナールを開いたか

私は1978年から12年間を、ヤマギワという照明会社の研究所に在籍して様々な照明デザインとそれに関連する仕事の現場を経験した。この研究所時代に建築照明デザインに没頭した私は、幸運にも建築家の磯崎新、槇文彦、原広司、伊東豊雄、安藤忠雄などという偉大な方々の仕事に協力する機会をたくさん得た。研究所の12年間は充実した時間の積み重ねだった。今の私の基礎体力や余剰筋力はこの時代に鍛えられたものである。そして私は1990年にTLヤマギワ研究所を退職し、日本で初めて本格的な建築照明デザイン事務所となる(株)ライティング プランナーズ アソシエーツ（通称LPA）という会社を6名の同志とともに設立したが、そのLPAも既に22年を経過し、多くの有能な競合他社を生みながら、東京、シンガポール、香港というアジアの主要都市にグループ会社を置き、50名もの熱く有能な社員に恵まれて多忙な日々を送るに至った。あっという間のヤマギワ研究時の12年間とLPAの22年間である。

災難と思える事態が2001年の春に起こった。いつの間にか私の兄貴分のような存在になっていたインテリアデザイナーの杉本貴志さんからの武蔵美への召集令状が届けられてしまったのである。もちろん、当時は大学の常勤教授を務めることなど考えられないほど多忙で充実した日々だったので、丁重に

何回もお断りした。数えてみると、当時私は11の大学で非常勤講師として単発的講義をしていたが、ほとんどは年に1回とか数回のものだった。それと常勤教授では責任の重さも全く違う。私にはできないので、5年ほど待ってほしいと杉本さんに何回も懇願したが、彼は自説を貫き通し私を口説き落としてしまった。

　最終的に私が武蔵美に来ることを英断したのは、杉本さんの熱意もさることながら、学生を教えることはデザインすることに類似していて、それに興味を持ったからだった。建築照明デザインの現場では、私が努力し汗水流すことによって、こんなに素晴らしい光環境が出現した…、という感慨だけが勲章だ。学生を教える意味もそこにある。明らかに私が学生のそばにいたからこそ現在の彼らが輝いて見える。そんな瞬間に立ち会いたいと思った。教育というのもデザインの方法論の1つではないのか。まして『光のゼミナール』など聞いたこともなく、光や陰影が私と学生との距離をどのように縮めてくれるのかにも、不安以上の期待を持ってしまった。最終的には、まあどうにかなる、始めてみようと思った。

光のゼミナールで何を目指したか

さて、光のゼミナールを開講するにあたって、どのような志や具体的な目標を設定すべきなのだろうか。このことは常に10年間を計画する癖のついている私には自分に出した初めの課題のようなものだった。私は既に世界的にも著名な照明デザイナーの一人として数えられ、ストックホルムのKTHや、ドイツのウィスマール工科大学、そしてNYCのパーソンズなど、世界の照明デザイン教育の筆頭にあげられる教育機関にも招聘されて照明の講義を行うことがあった。さあ、武蔵美の空間演出デザイン学科という器の中、私の多忙なデザイナーとしての仕事に両立させて、どのようなことができるのだろうか。これが私に課せられた条件であり課題だ。

　私は当初、大学院レベルだけで専門的な光のデザインのプロを目指す教育をしたいと考えていた。世界一の光の専門的なゼミになる夢を捨てきれなかったからである。しかし、現実的には学部生を教えるとなると、彼らは照明の基礎知識さえも持ち合わせていない。所詮、光の専門的な講座を開講するにはリスクが大きすぎる。大半の学生は「照明デザイナーになりたい訳ではないが光に興味がある…」というモチベーションだ。私は早々に方針を切り替えて「光のデザインを通じて多彩なデザイン分野を横断するゼミナール」を発想した。もともと空間演出デザイン学科はファッション、セノグラフ、インテリア、環境デザイン、というような多彩なコースと教授陣で構成されているので、これらすべての領域を光のデザインで串刺しするような姿勢がふさわしいのではないかと結論付けた。

　そもそも光のデザインとは、情報量の87%を占めるとされる視覚情報の品質や操作に関わって成立するものだ。目を閉じてしまわない限り、昼夜に渡り私たちの周りには無数の景色や視覚情報が飛び込んでくる。それらの情報のあり方を丁寧に紐解いていくのが光のデザインの役割であり、光のデザインを駆使することによって私たちの日常が数倍も良質になることを夢見ているのである。であるから、あらゆるデザインを志す者はすべからく、まずは光のデザインを学ぶべきなのだ。

　そのように考えると、急に肩から力が抜けて自然体の私が見えてきて嬉しかった。学生を教えることを通じて、私自身が鍛えられ成長することにも興味を覚えた。照明デザインという日常の仕事の中では発見しにくいことが、光のゼミナールでは感じられそうな気がしてきた。ここでは常に光という素材を媒体としながらも、学生と一緒になって様々なデザインの対象を探して行こうと考えた。光というフィルターを透かして社会を俯瞰する。そのくらいの方が美術大学らしくて恰好良いとも思えた。

ゼミ課題と課外プロジェクトの役割

武蔵美で光のデザインを教えるにあたって、他大学の教育プログラムを参考にすることはなかった。世界中を探しても照明デザインの講座を持つ機関は少なく、そのほとんどが電気設備工学または照明光学的なもの、または環境知覚心理学的なものを教えるか、あるいは正反対に技術論は無視してアート感覚

的な作品ばかりを作らせるか、に偏っている。

　私は光のデザインは科学と芸術の間を行ったり来たりするものだと信じているので、極端に技術寄りになったり、芸術家を気取ったりしたくないと考えていた。デザイナーというのは科学と芸術と経済とをシャッフルしながら、その渦の中で回答するものだ。だから大学教育の中で定められたシラバスに掲載するような教育プログラムは、光のゼミナールで必要かろうと結論付けた。しかし、何か光のゼミナールを示すミッションのようなものが必要だと考えて、学生には以下のような文言を配布した。

　　　面出ゼミは、光をキーワードにして「人 - モノ - 空間 - 環境」の関係を研究学習し、新たな生活情景をデザインするためのフォーラムです。ゼミの内容は三種類の指導方針によって構成されます。
　　1 光の調査分析
　　　光の調査分析を通じて、人と光の関係を探る。
　　　ライトコレクション、照明探偵団への活動参加
　　2 光の実験演習
　　　光の実験演習を通じて新たな光と環境に出会う。
　　　ライトアップゲリラ
　　　キャンドルナイト@ Omotesando Eco-Avenue
　　3 光の創造
　　　光のデザインを通じて新しい人間生活を提案する。
　　　個人制作課題 / 課題に革新的に解答する。
　　　卒業制作 / 自己を強烈に表現する。

　私はデザイン教育というのは、まず自らの感覚機能を鍛え直すことから始まるべきだと考えている。対象を鋭く正確に観察し、そこから何かを読み取る力を養わねばならない。とりわけ光や音や匂いに関しては、現代人は江戸時代以降、社会が生活の利便性だけを優先するにつれて、視覚、聴覚、嗅覚の機能を退化させてきた。明る過ぎる日常空間、大き過ぎる音、過剰な匂いの演出などが私たちの知覚機能を鈍化させてきた。それをまず初めに取り戻さねばならない。だからゼミの初めには自然光や都市光をひたすら見て歩き、光を採集する行為を続ける。それと並行してゼミ生は照明探偵街の歩きなどにも引き込まれ、じっとしている暇がないほどフィールドに出る。

　調査分析を繰り返した後には、やはり巷に出て実際の光を使って光の実験演習に駆り出される。武蔵美には舞台照明用の器具がたくさん設備されているので、それを現場に運びアノニマスな都市空間を勝手にライトアップして叱られないうちに足早に逃げてくる、という課題を行う。それをライトアップ・ゲリラ（さっと照らして、さっと逃げろ！）と呼ぶことにした。毎年主体的に参加したキャンドルナイト@ Omotesando Eco-Avenue というイベントでは、キャンドルの火を使った不自由ではあるがいとおしいあかりの実験演習をする。東京の表参道ケヤキ道を舞台として、キャンドル・インスタレーション、キャンドル・カフェ、神宮前小学校の子どもたちとのあかりワークショップ、オリジナル行灯を製作しパレードを行う。この一連のパフォーマンスがキャンドルという僅かな光による挑戦だ。ライトアップ・ゲリラという環境照明と、キャンドルの僅かなあかりを頼りにしたパフォーマンス。この両者が光の実験演習のコアになっていった。

　光の調査分析を行い、実験演習を経過した後になってやっと、光の創造という個人に向けたデザイン課題を出す。これは主に4年になっての前期に集中することが多いが、4年後期が卒業制作1課題に当てられることもあって、個人の創造力を存分に発揮する機会を与えたいと思った。出題の意図は単純明快に示し、各自が自分の個性を生かして自由に解釈展開できるように考えている。予備校時代から教えられることに従順な学生は、自由に自分のテーマを設定して課題をこなしていくことが苦手である。自分を探し当てること。自分の進路や方向性を潔く決定して行くこと。それらは常について回るデザイナーの宿命だが、卒業制作をどう克服していくかが光のゼミナールの最終課題であることは間違いない。そこにはほとんどの場合に100%の成功例は見られないが、それ以上に役立つ失敗例には心打たれることもある。

　教育プログラムというほどのものではないが、光のゼミナールの実態を示す中心的課題が浮き彫りにされてきた。ゼミの10年間をまとめるという今回の編集意図に従って作業した成果でもある。ゼミOB生

を含んで数回のワークショップを行い、「光のデザイン、7つの学び方」なるキーワードを抽出した。見ることから始まって新たな光を創造するまでの学生に要求される所作を7項目に分けて本書では詳細に紹介することにした。光を観る、光を集める、光に触れる、自然光に学ぶ、闇を知る、影と遊ぶ、光を操る。この7種類の行為を私は常に学生に要求した。違ういい方をすると、これらの学習態度が光のデザインを学ぶための基礎要素なのである。自然科学や社会科学の一部として学ぶことのできる照明デザインとは別に、光のデザインという所作を学ぶために不可欠な要素であると確信する。

7つの要素の学習成果はいつも曖昧であり、明確な評価対象には成り得ないことも多いが、やり方を真似て繰り返すだけのことでも十分な意味を持っている。目を大きく見開いて穴の開くほど対象を観る訓練、耳の穴を穿って僅かな音を聞き分ける訓練、そのようなものは生涯続けなければならない。それだから学生は光のゼミナールを通じて、生涯教育のプログラムの入り口を体験したことになるのかもしれない。私は学生を卒業して40年近くにもなるが、私自身この7つのプログラム要素を未だに楽しく実践することができている。

デザイン教育と今日的課題

十年一昔というが、思い返してみるとこの10年の社会変化はそれまでの10年と比較のしようのないほど劇的なものであった。10年前に日本経済のバブルははじけていたけれど、リーマンショックに始まる世界経済危機の勃発、中国市場の台頭、地球温暖化による異常気象、アメリカに次いで欧州経済の破壊、異常な円高と日本経済の低迷、LEDの急台頭、モバイル時代の到来、東日本大災害と今なお続く原発放射能災害など。どう見ても明るい話題に乏しい10年としかいいようがない。

しかし、この閉塞感を伴う10年の出来事は「予想を超える」突発的な出来事ではない。賢明な判断を欠いたことのツケが回ってきただけではないだろうか。これらのツケが教育の現場にも回ってきた。少子化の波による学生不足もその1つで、そもそも過剰な数を誇っていた日本の大学は規模の縮小や合併、そして廃校を迫られる事態を迎えた。美術大学の受験性も年々その数が減少し、美術大学の社会的な役割の変化や、社会経済に連鎖した教育制度のあり方が問われている。テクノロジーと芸術文化の課題、コミュニケーションの進化と退化、デザイン新領域と社会的役割の変化、少子化の波と大学のビジネス化など。武蔵美の社会的な価値や空間演出デザイン学科の役割は、どのように変化していくべきなのだろうか。来年再来年という話ではなく、少なくても10年20年という視野と展望を持って、建設的に変化再生を期してもらいたい。

一方、美術大学で学ぶ学生の体質変化も著しい。武蔵美には押しなべて真面目で従順な学生が多いという印象だが、それはアバンギャルドを旨とすべき美大にあっては必ずしも安堵すべき傾向ではない。真面目で従順は、家族に手を掛けられて育ったひ弱な果実を連想させる。しかも急激なIT時代に幼少を過ごした体験は、PCやモバイルホーンなどへの愛着は強いが、目と目を合わせたコミュニケーションが苦手という現象を多発させている。幸い私のゼミ生は一緒に楽しい酒も飲むし、喜怒哀楽を共にして、積極的に打ち解けてくれるが、一般的には学生の酒量も減って、馬鹿をやり自分をさらけ出す美大生が少なくなっている。人前で恥をかきたくないなどという学生は即座に更生させねばならない。早くたくさん恥をかくことこそ学生の特権であるからだ。

10年間の光のゼミナールは、新しいデザイン領域を目指す実験土俵でもあったと思う。私はその過程や成果に必ずしも満足をしてはいないが、私たちの挑戦が失敗したとは思っていない。同時代を共有した職場としての大学の仲間たち、そして私のゼミを卒業した140名ほどの学生諸君にとって、この一冊の本が想い出アルバム以上の役割を果たせたら幸いである。

Preface: The Light Seminar
Ten Years of Experiments

It is nearly 35 years since I first became involved in the design of light. Ironically, when I decided to become a lighting designer in 1988, I felt there was something dubious about the profession. With all due respect to the leaders of the profession at the time, lighting design then was synonymous with magnificence and glitter and meant the design of household lighting fixtures--products displayed in storefronts. Though I had studied product design at a university, it seemed to me ultimately a way of producing a mountain of industrial waste; I wanted to avoid it.

I therefore find it difficult even today to say why I became a lighting designer. There frankly wasn't a good reason. I had no choice because I could not find any work in environmental design, the field in which I was interested. This does not make for a very stirring account. When I did enter lighting design, the world of which I had been critical, I called it "the design of light." It felt better to think of it as the design of a light environment. I would design, not the look or the shape of lighting fixtures, but the profound relationship between light and darkness and people. It would be a challenging world of limitless possibilities. I foresaw a delightful yet difficult future for the design of light.

How I Learned to Design Light

I had not previously studied the design of light. It was at the research center of a lighting company that I began to study the design of light for the first time. On learning that architectural lighting design was already an established genre in the eastern United States, I made a deliberate effort to go out into the field to experience both natural light and the effect of light on architectural spaces throughout the world. The intensive growth of the Japanese economy made it possible. The research center for the lighting company sent me off to various countries to undertake studies because I was relatively proficient in English. I pored over architectural books and magazines, analyzed lighting catalogs from overseas and studied technical papers edited by the Illuminating Engineering Institute of Japan. However, I did not trust theories unreservedly. I was persuaded by theoretical explanations only when they were borne out by my own experience with light, and at times I rejected such explanations altogether. I was frankly an empiricist. The more new things I saw, the more aspects of light I found myself assimilating through my eyes, brain and body. I learned to value that sensation. I had a chance to collaborate with a lighting consultant with a busy practice on the American East Coast, and imitated what I admired about his movements and technique. I visited and was moved by many modernist works of architectural illumination in the United States by architects such as Eero Saarinen, Mies van der Rohe, Louis Kahn, Philip Johnson and I. M. Pei. I also learned that light in a historic, premodern building, even a Japanese sukiya-style building, Shinto shrine or Buddhist temple, represented an attempt to design a light environment within a particular cultural context. Everything I saw was a part of a process of learning how to design light.

The Reason I Began the Light Seminar

For 12 years, I was a member of the research center of a lighting company called Yamagiwa and in that capacity dealt with a number of project sites. During that period, I was absorbed in the design of architectural lighting and fortunately had many opportunities to cooperate with outstanding architects such as Arata Isozaki, Fumihiko Maki, Hiroshi Hara, Toyo Ito and Tadao Ando. I had many fulfilling experiences during those years at the research center. I built up my stamina and reserves of strength in that period. In 1990 I retired from TL Yamagiwa Laboratory and with six colleagues established the first genuine architectural lighting design office in Japan, a company called Lighting Planners Associates (LPA). Now in its 22nd year, LPA has given rise to many capable rival companies. It maintains offices in Tokyo, Singapore and Hong Kong and has a staff of 50 enthusiastic, talented individuals. The years at Yamagiwa and with LPA have been busy and flown past.

I met with "misfortune" in spring of 2001. Takashi Sugimoto, the interior designer, who had become like a big brother to me, told me to report for duty to Musashino Art University. My days were so busy and fulfilling at the time that being a full-time professor at a university seemed unthinkable. I politely refused, not once but several times. I was then a part-time lecturer at eleven universities, but that mostly meant giving a few lectures, sometimes just one lecture, a year at any given institution. Being a full-time professor would mean an entirely different level of responsibility. Since I could not do it, I repeatedly asked Mr. Sugimoto to wait about five years. However, he refused to compromise and eventually got his way.

What ultimately made me decide to come to Musashino Art University was not only Mr. Sugimoto's zeal, but the fact that I myself had long been interested in the similarity between teaching students and designing. On the site of an architectural lighting design project, the only badge of honor one receives is the joy that comes from having produced a wonderful light environment through one's efforts. Teaching students is much like that. I wanted to feel the elation that comes from having played a role in whatever they accomplish. Education is after all a methodology of design. Moreover, I was intrigued by the idea of a "light seminar," something with which I was unfamiliar. It was with both anxiety and anticipation that I wondered how I might establish a relationship with students on the basis of light and shadow. In the end, I thought, it will work out somehow, so why not begin?

The Aim of the Light Seminar

I now had to decide what sort of aim or specific objective to set for the Light Seminar. Being in the habit of making ten-year plans, I made this my first task. I already had an international reputation as a lighting designer and was invited to lecture on lighting at leading educational institutions in the field of lighting design throughout the world such as KTH in Stockholm, Hochschule Wismar in Germany, and Parsons in New York City. How would I be able to teach within the Department of Scenography, Display and Fashion Design in Musashino Art University and continue my busy practice as a designer? For those were the conditions and the task I was given.

Initially, I wanted to teach only students at the graduate school level who were aiming to become professional designers of light. That was because I dreamed of creating the top technical seminar on light in the world. However, the reality was that I would be teaching undergraduate students, and they would not have even basic knowledge about lighting. There was too great a risk in teaching them a technical course on light. The majority of students would be motivated by curiosity about light rather than desire to become a lighting designer. I quickly decided to change my plan and conceived the idea of a seminar that would touch on various fields of design through the design of light. The department had a diverse faculty and a set of courses in fashion, scenography, interior design and environmental design to begin with, so I concluded that the proper approach to take would be to knit all these fields together through the design of light.

The design of light basically has to do with the quality and manipulation of visual information, which is said to account for 87 percent of all information. Unless one closes one's eyes, one takes in an endless stream of visual information from the immediate environment, day and night. Carefully unraveling these diverse strands of information is the role of the design of light, and the dream is to improve the quality of our everyday lives many times over through the use of the design of light. That being so, a person who intends to enter any field of design ought to learn the design of light first.

Having developed this line of reasoning, I suddenly relaxed and was happy. I was intrigued by the idea that by training students I myself would grow and become more disciplined. I felt that in the Light Seminar I might experience things that are difficult to find in the everyday work of lighting design. Using light as a medium, I would search together with students for diverse targets of design. Through a filter of light, we would get an overview of society. I thought, that would be a goal suitable for an art university.

Educational Program:
The Role of Seminar Assignments and Extracurricular Projects

In teaching the design of light at Musashino Art University, I did not reference educational programs at other universities. There are only a few institutions anywhere in the world with a course in lighting design. Most lean one way or the other, either teaching only the technical aspects of light—that is, electrical systems, illumination optics or environmental perceptual psychology—or having students take an artistic approach to the subject and ignoring the technical aspects. Convinced that the design of light was a matter of going back and forth between science and art, I did not want to adopt an extreme, technically-oriented approach or an approach that was oriented too much toward art. A designer shuttles between science, art and economics and arrives at a solution somewhere in that mix. I therefore concluded that the Light Seminar would not need an educational program of the sort described in an established university syllabus. Nonetheless, I distributed the following text to students in the belief that something like a statement of the mission of the Light Seminar was needed.

> *The Mende seminar is a forum for studying the relationship between human beings, space and the environment and designing new 'scenes of life,' with light as the keyword. Guidance will be provided in three ways, and these will dictate the way the subject matter covered by the seminar is organized.*
>
> *1 Study and analysis of light: the relationship between human beings and light will be explored through the study and analysis of light.*
> *Light collection and participation in the activities of the Lighting Detectives.*
> *2 Experiments and exercises with light: new lights and environments will be encountered through experiments and exercises with light.*
> *Light Up Guerrilla, Candle Night@Omotesando Eco-Avenue.*
> *3 Creation of Light: new ways of living will be proposed through the design of light.*
> *Individual work assignments: innovative answers are to be presented to given assignments.*
> *Graduation work: each student is expected to produce a work of intense self-expression.*

I believe a design education ought to start with the re-disciplining of one's own sensory functions. One must develop the ability to observe a subject sensitively and accurately and to grasp what one is seeing. Especially with respect to light, sound, and smell, we have allowed our senses of vision, hearing and smell to atrophy since the Edo period because society has placed so much priority on making life more convenient. We must first recover those senses. Therefore, at the beginning the participants in the seminar do nothing but walk about looking at light, both natural and urban, and collecting specimens of light. Students also take part in walks organized by the Lighting Detectives. As a result seminar students are busy being out in the field.

After repeated surveys and analyses, students go out into the field once more, this time to undertake experiments and exercises using actual light. There are plenty of stage lighting equipment at Musashino Art University; the assignment is for students to carry such equipment out into the field and light up anonymous urban spaces and get away before they are chastised. I called this "Light Up Guerrilla." (The trick is to quickly light up a scene and quickly flee.) The event called Candle Night@Omotesando Eco-Avenue, of which I am one of the main organizers each year, is a chance to carry out experiments and exercises with candlelight, which can be inconvenient but charming. The stage is the zelkova-lined street known as Omotesando in Tokyo; there, installations of candles are set up, candles are used for lighting cafés, a light workshop is held with children from the Jingumae Elementary School, original lanterns are prepared and a parade is held. This series of performances is an opportunity to take up the challenge of using the faint light from candles. Attempts at environmental illumination that I call Light Up Guerrilla and the performances that depend on the faint light of candles are the core experiments and exercises using light.

After students have carried out surveys and analyses as well as experiments and exercises, I at last give out design exercises where individuals must create light. These tend to come in the first semester of their fourth year, but the second semester of the fourth year can be given over to a single assignment constituting the graduation work. I wanted to give students plenty of opportunities to do creative work individually. The aim of an assignment is indicated simply and clearly and conceived so that each student is free to interpret and develop the project and give full play to his or her own personality. Students for whom education has been a matter of simply accepting what is taught, ever since their days in cram school, find it difficult to decide on a theme and finish an assignment on their own. To discover oneself and boldly decide one's own course or direction is something that is always required of a designer. Without a doubt, the final assignment in the Light Seminar is to overcome the difficulties presented by the preparation of the graduation work. A work that is 100 percent successful is rare, but I am sometimes even more excited by examples of useful failure.

It may be an exaggeration to call it an educational program, but the main issues actually addressed in the seminar have been made clear after the ten years of the seminar were reviewed for the purposes of this publication. A workshop that included past students of the seminar was held and identified keywords that suggested "Seven Ways to Study the Design of LIght." It was decided to detail in this book seven courses of action I demanded of students, from looking to creating new light. See light, gather light, touch light, study natural light, become acquainted with darkness, play with shadows and manipulate light. I always required students to pursue these seven courses of action. To put it another way, these are attitudes to study that I believe are fundamental to learning how to design light. I am convinced that these are indispensable to the study of the design of light as opposed to lighting design, which can be learned as a part of natural or social science.

The results of studying those seven elements are always ambiguous and often impossible to evaluate clearly. However, simply imitating and repeatedly adopting this approach is sufficiently meaningful. Disciplining oneself to open one's eyes as wide as possible and observe the subject or to listen carefully and distinguish between faint sounds--one must continue to do such things throughout life. For a student, therefore, the Light Seminar may be the first step in a lifelong education program. It has been nearly 40 years since I was a student, but I still enjoy practicing these seven program elements.

Design Education and Contemporary Issues

They say ten years can produce a sea change, and indeed the social changes that have taken place in the last decade have been unprecedented and dramatic. The Japanese economic bubble had already burst ten years ago, but in the last ten years we have experienced the world financial crisis that began with the bankruptcy of Lehman Brothers, emergence of China as an economic power, meteorological anomalies caused by global warming, failure of the European economy, extraordinary rise in value of the yen, economic slump in Japan, emergence of LEDs, advent of the age of mobile devices, Great East Japan Earthquake and still unresolved problems of radiation from the nuclear power plant. There has not been much to cheer about in the last decade.

However, the events of the last ten years, which have indeed given rise to a besieged mentality on the part of many Japanese, were not totally unexpected or sudden. We are simply paying the price for having made unwise decisions in the past. We are paying the price for unwise decisions in the arena of education as well. The shortage of students arising from a dwindling birthrate is one such problem, and Japanese universities of which there was an excess are now having to downsize, merge or close down altogether. Applicants to art universities are decreasing year by year, forcing those institutions to consider changes in their role in society or to question the way the educational system is linked to society and the economy. Issues of technology and artistic culture, the development and degeneration of communication, new areas of design and changes in social function, the declining birthrate and the transformation of universities into businesses. How should the social value of Musashino Art University and the role of the Department of Scenography, Display and Fashion Design change? I hope that constructive change and reform will take place with a view toward, not the next year or the year after that, but at least 10 or 20 years into the future.

There has also been a striking change in the character of students at art universities. Students at Musashino Art University all seem serious and obedient, but that is by no means a good thing at an art university, a place that ought to embrace the avant-garde. The image of delicate, hot-house flowers springs to mind. Moreover, the experience of having been a child during the rapidly developing age of IT has led to not just a strong attachment to PCs and mobile phones but also frequent difficulties with face-to-face communication. Fortunately, the students in my seminar enjoy having a drink together, readily share their feelings and generally deal with each other in an open way. However, students in general do not drink as much as they used to, and fewer art students make fools of themselves nowadays. Students who say they do not want to make fools of themselves in public need to be rehabilitated immediately. Making a fool of oneself quickly and often ought to be the prerogative of the student.

The ten years of the Light Seminar also represented a testing ground for developing a new field of design. I am by no means satisfied with the process or the results but do not believe our efforts were in vain. I hope that this book will serve as more than just a memento for my university colleagues and the approximately 140 students who completed my seminar.

Kaoru Mende

光のデザイン ── 7つの学び方
Seven ways to study the design of light

光のゼミナールが費やしたエネルギーや成果を、ゼミ課題やゼミプロジェクトを中心にカード上に記録して、それぞれが何を目指していたのかを分析した。その結果、デザインは最終的に創造的何かを生み出す行為ではあるが、面出ゼミでは一般的なデザイン課題が少ないことが判明した。

　光のデザインはどんなことに拘って学習するのか。私たちは頻繁に発せられる言葉の中から7つの学び方を抽出した。光を観る、光を集める、光に触れる、自然光に学ぶ、闇を知る、影と遊ぶ、光を操る。この7つの学び方は光のゼミナールで反復されたキーワードでもある。

The effort put into the Light Seminar and the fruits of that effort, mainly in the form of seminar assignments and projects, were recorded on cards, and the aims of those assignments and projects were analyzed. It was discovered as a result that, though design is ultimately a creative act, there were few general design assignments given in the Mende seminar.

　What sorts of things were studied in order to learn how to design light? We identified seven ways to study light from often used phrases: see light, gather light, touch light, study natural light, become acquainted with darkness, play with shadows, and manipulate light. These are keywords that were often repeated in the Light Seminar.

光を観る
Observing Light

先ずは見る、視る、観る…
健脚にものをいわせて無防備に
光を観る
芸術的な光に大袈裟に感動する
嫌な光に憤慨する
感動と憤慨の理由を推理する

First, see, view, observe, etc.
Observe light defenselessly while
vigorously walking
Be overly thrilled by artistic light
Be outraged by distasteful light
Deduce the reasons for the thrills or outrage

光を集める
Collecting Light

視野内の景色から光を切り取る
光と影に接近する
光採集（Light collection）
＝光の標本づくり
光の英雄と犯罪者を見極める
詩情を持って観た光を語る

Extract light from a landscape in sight
Get close to light and shadow
Collect light, or sample light
Determine whether the light examples are
heroes or villains
Talk poetically about the light you observed

光に触れる
Touching Light

光は嘘をつかない
光はイメージ通りの仕事をしない
時に光に触れて火傷をする
そして光に感動する
何度も失敗しながら
光と親友になる

Light never tells a lie
Light never works like its image
Touch light and get burned
from time to time
Be thrilled by light
Develop a friendship with light
through innumerable failures

自然光に学ぶ
Learning from Natural Light

自然光と共に生きる
自然光とは太陽の光と火のあかり
全ての感動が自然光の中にある
自然光の技を丁寧に盗む
色温度、光の位置、陰影の深さ、
移ろいの速度など…

Live with natural light
Natural light is composed of solar light
and the light of the fire
All thrills exist in natural light
Carefully steal the art of natural light
Color temperature, location of light, depth
of shadow, velocity of changing light, etc.

闇を知る
Knowing Darkness

先ずは闇から始まる
闇を恐れる
闇を知り闇を受け入れる
闇なしで光の美学は成立しない
闇は美しいか？

Everything starts in darkness
Fear darkness
Know and accept darkness
No aesthetics of light is formed
without darkness
Is darkness beautiful?

影と遊ぶ
Playing with Shadows

世界は様々な陰影でできている
僅かな光にこそ快い影が宿る
月影そしてローソクの灯が
創る影
影と戯れる
影を通じて光を知ろう

The world is made of various shadows
Dim light makes comfortable shadows
Moon's shadow and shadows made
by candlelight
Play with shadows
Know light through shadows

光を操る
Manipulating Light

いよいよ光のデザインを行う
内なる抑圧とイマジネーションを
爆発させる
光と影を自由に創作する
住環境、商業環境、都市環境…
さあ、光を操れ、陰影を操れ

Finally, design light
Blast your inner oppressiveness and
imagination
Freely create light and shadow
Residential, commercial, urban
and other environments
Manipulate light, manipulate shadows

光を観る
OBSERVING LIGHT

光を観る
OBSERVING LIGHT

光のデザインの学習法の基本となるのは、現場で生の光をつぶさに観るという行為だ。私もこの基本動作を35年間、欠かさず毎日行っている。光は日常のあらゆる場面に散らばっている。先入観なしに現場で様々な光に遭遇し、その光の様子を克明に評価し記録する。

　光を観るとは、見る、視る、と書くこともあるが、目で見るのみでなく網膜上の視覚情報は脳で感知し、評価判断し、記憶するというプロセスで成立するので、同じ現場で同じ光を見た場合にも、人により受け取り方や評価が異なることが頻繁に起こる。美しい光景をインターネットで探す学生が大勢いるが、これは光のデザインの学習法としては全くいただけない。バーチャルな光に真実はないからだ。現場に立って、視覚、聴覚、嗅覚、時には味覚や触覚まで使って光を観る必要がある。光の感覚は常にリアルで生々しく個人的であるべきだ。光のデザインには想像力でなく、観察力や感知力を鍛える必要がある。

　私は照明探偵団という非営利の実践的照明文化研究会を20年以上も運営しているが、これの基本も先入観なく光の現場を見て歩くことだ。極端に無防備になって、行き当たりばったりの様々な光の現場を徒党を組んで歩き回る。照明探偵団の街歩きにはもちろん学生も参加する。巷には光の事件がたくさん潜んでいるので、それに深く感激したり、ひどく憤慨してみたりする。これはゲームみたいなものであるが、いい音楽をたくさん聴くと耳が肥えたり、美味しい料理を食すると味覚が鍛えられるように、たくさんの種類の光や陰影の現場を体験するうちに、光に対する知覚機能や判断基準もグレードアップしてくるから面白い。

　2003年からは世界照明探偵団（TNT）フォーラムという、年に一度の海外で行われるワークショップやシンポジウムにゼミ参加を続けた。ストックホルム、ハンブルグ、ニューヨーク、シンガポール、コペンハーゲン、ベオグラード、北京など。ゼミ学生を海外の光環境調査に連れ出すと、日本で目にする常識的な光も他国では非常識に変化することもあることを知る。異なる歴史や文化を持った都市に赴くと一層学び取るものも多い。光の現場は細かい差異の積み重ねでできている。

If one wants to learn how to design light, one must first go out into the field and carefully observe light. This is something I myself have done every day for 35 years. Light is scattered in every conceivable scene of everyday life. One needs to encounter diverse forms of light without any preconceptions and evaluate and record their appearance in detail. To observe light is not simply to see light with one's eyes; the brain senses the perceptual information received on the retina and evaluates, judges and records that information. The same light in the same scene can often be interpreted and evaluated differently, depending on the person. Many students search the Internet for beautiful scenes, but that is a poor way to learn how to design light. That is because there is no truth in virtual light. One needs to stand out in the field and to observe light using the senses of vision, hearing, smell and at times even taste and touch. The perception of light must always be real, vivid and personal. In order to design light, one must improve one's powers of observation and perception rather than one's power of imagination.

　For over 20 years, I have been running an organization called the Lighting Detectives (Shomei Tanteidan), a non-profit society whose objective is to study the practical aspects of the culture of lighting. It too is based on the idea of looking at light out in the field without preconceptions. Receptive to stimuli and without any plans, members band together and walk about, looking at light in different places. On such walks in town, the Lighting Detectives are accompanied by students. There are many things concerning light in the busier parts of town that impress or anger us. It is like a game. The more one listens to good music, the more sensitive one's hearing becomes, and the more one eats delicious food, the more sensitive one's taste becomes. Similarly, the more one experiences different types of light and shadow, the better one's perception and standard of judgment regarding light become.

　Since 2003, the seminar has taken part in an annual overseas workshop or symposium called the Transnational Tanteidan (TNT) Forum. It has been held in cities such as Stockholm, Hamburg, New York, Singapore, Copenhagen, Belgrade and Beijing. About 15 seminar students are taken abroad on these occasions to survey light environments; they discover that a light that is acceptable in Japan can turn out to be unacceptable in another country. There is a great deal one can learn, particularly on a visit to a city with a different history or culture. Light, out in the field, is made up of many layers of subtle differences.

世界の光を観る
Observing lights all over the world

何のための光の世界旅行だったか…
What was the purpose of the trips to other countries for experiencing light?

偶然にも大学に赴任した2002年から、私が主宰する世界照明探偵団フォーラムを開始した。年に一度の海外での光環境調査やワークショップ、これに学生を参加させることを直ぐに思いついた。日本を離れた世界で起きていることを一緒に体験させたかったからだ。

　光を観るにも同じようなものばかりを見てはいけない。極力、種類の違うものを対比させながら観察思考することで、学生はその差異を知り、戸惑い、それぞれの価値を探ることになる。だから私は許される範囲でできるだけ彼らを国外に脱出させようとした。

　学生たちは言葉の不自由さを嘆きながらも人々や街、そして食べ物や光環境の相違をこと細かく感じ取っている様子だった。文化を比較体験することに尽きる。

I happened to launch the forums of the Trans-national Lighting Detectives as the head in 2002 when I took a post in the university. I instantly got an idea of having students participate in annual overseas light environment surveys and workshops. I wanted them to experience what was happening in other countries. You should not exclusively observe similar types of light. Students might know the difference, get bewildered and seek the value by comparatively observing and considering different types of light inasmuch as possible. I therefore tried to send them to other countries as long as allowed.

　Students seemed to be feeling minute difference in people, city, food and light environment while they grumbled about the difficulty in verbal communication. Cultures need to be experienced in comparative perspective.

1	2	3	4
5	6	7	8
9	10		
11	12	13	14

1 ストックホルム/TNTフォーラムに参加して	1 Stockholm / Attending the TNT-Forum
2 バリのワークショップ/巨大な光の凧揚げ	2 Bali Workshop / A giant luminous kite
3 バリのワークショップ/凧揚げ会場でガムランの演奏	3 Bali Workshop / Watching a gamelan performance at the kite festival
4,5 シンガポールの街角にて	4,5 On a street corner in Singapore
6 マレーシアの街角にて	6 On the streets of a Malaysian city
7 パリの街角/地中埋設器具の前に立つ	7 Streets of Paris / Standing before buried fixtures
8 パリのシャンゼリゼ大通り	8 Le Champs Elysees in Paris
9 コペンハーゲン/ストロイエの夜景調査	9 Copenhagen / Night survey on Stroget
10 コペンハーゲンではレンタサイクルでの移動	10 Rent-a-cycle, our means of transportation in Copenhagen
11 コペンハーゲン/クリスマスツリーの前で	11 Posing before a large Christmas tree in Copenhagen
12 ベオグラードの夜景調査	12 Night survey in Belgrade
13 北京/万里の長城	13 Beijing / The Great Wall of China
14 北京/天安門広場にたつ	14 Beijing / Tiananmen Square

光を観る
OBSERVING LIGHT

Stockholm, SWEDEN
August 2003

Copenhagen, DENMARK
December 2007

Beijing, CHINA
October 2009

Belgrade, SERBIA
September 2008

Hamburg, GERMANY
September 2004

Bangkok, THAILAND
March 2012

SINGAPORE
November 2006

Tokyo, 2002   *Stockholm,, 2003*   *Hamburg, 2004*   *New York City, 2005*

Tokyo, JAPAN
September 2005

New York City, U.S.A.
September 2005

世界照明探偵団フォーラムの開催内容一覧　Contents table of TNT forums

	Place	Period	Theme	Symposium	Workshop	Tour	Exhibition
1	Tokyo, JAPAN	2002	Regional Lighting	*			*
2	Stockholm, SWEDEN	2003 08.29 – 31	Residential Neighborhood Lighting	*			*
3	Hamburg, GERMANY	2004 09.02 – 04	Daily Transportation Facilities	*			*
4	New York City, U.S.A.	2005 09.21 – 24	Main Street Lighting	*			*
5	SINGAPORE	2006 11.22 – 23	Façade Lighting	*	*	*	*
6	Copenhagen, DENMARK	2007 12.03 – 05	PARKS, PLAZA and PROMENADES	*			*
7	Belgrade, SERBIA	2008 09.16 – 19	Lighting Identity of Cities	*		*	*
8	Beijing, CHINA	2009 10. 13 – 16	Enjoy Lighting with Ecology	*	*	*	*
9	Bangkok, THAILAND	2012 03.01 – 03	Bangkok Lighting Identity	*		*	*

Singapore, 2006　　　*Copenhagen, 2007*　　　*Belgrade, 2008*　　　*Beijing, 2009*

照明探偵団とは何だ?
What is Lighting Detectives?

照明探偵団は1990年に東京で結成された実践的照明文化研究会である。発足以来、都市の光環境調査を継続的に行い、その営利を目的としないフィールドワークの成果を出版や展示会、その他のイベントを通して様々なかたちで発信している。その後、活動は照明の専門家から学生・主婦に至るまで広範な市民メンバーにまで拡大し、現在250名ほどの照明探偵団倶楽部会員を登録している。

国際的な照明探偵団活動への興味の高まりとともに、1999年には国境を越えて"世界照明探偵団/Transnational Lighting Detectives"がスタートし、現在各国のコアメンバー11名を中心に活動を行っている。

The Lighting Detectives is a society for the study of lighting culture formed in Tokyo 1990. Since its inception, it has continued to carry out studies of the light environment in cities and to make public the results of its non-profit field work in various forms through publications, exhibitions and other events. The society has since grown to include everyone from lighting experts to students and housewivesAt present there are 250 club members.

In response to growing interest abroad in the society's activities, the Transnational Lighting Detectives was formed in 1999. A network has been established around 11 core members from different countries to carry out various activities.

照明探偵団の目指すものは…
- 展望｜地域の照明文化を通じて、知識と感動を分かち合うためのプラットフォームを創りだす
- 使命｜比較文化としての照明のアイデンティティを楽しみ、尊重し、学習する
- 目標｜照明の未来を共に語りあう

活動内容
照明探偵団の活動は主に以下の7つが主軸となっている
1 日常的な街歩きとサロン活動
2 世界の都市照明調査
3 出版・メディア活動
4 セミナー・展示会の開催
5 市民参加型イベント
6 ニュースレターとウェブによる情報発信
7 世界照明探偵団

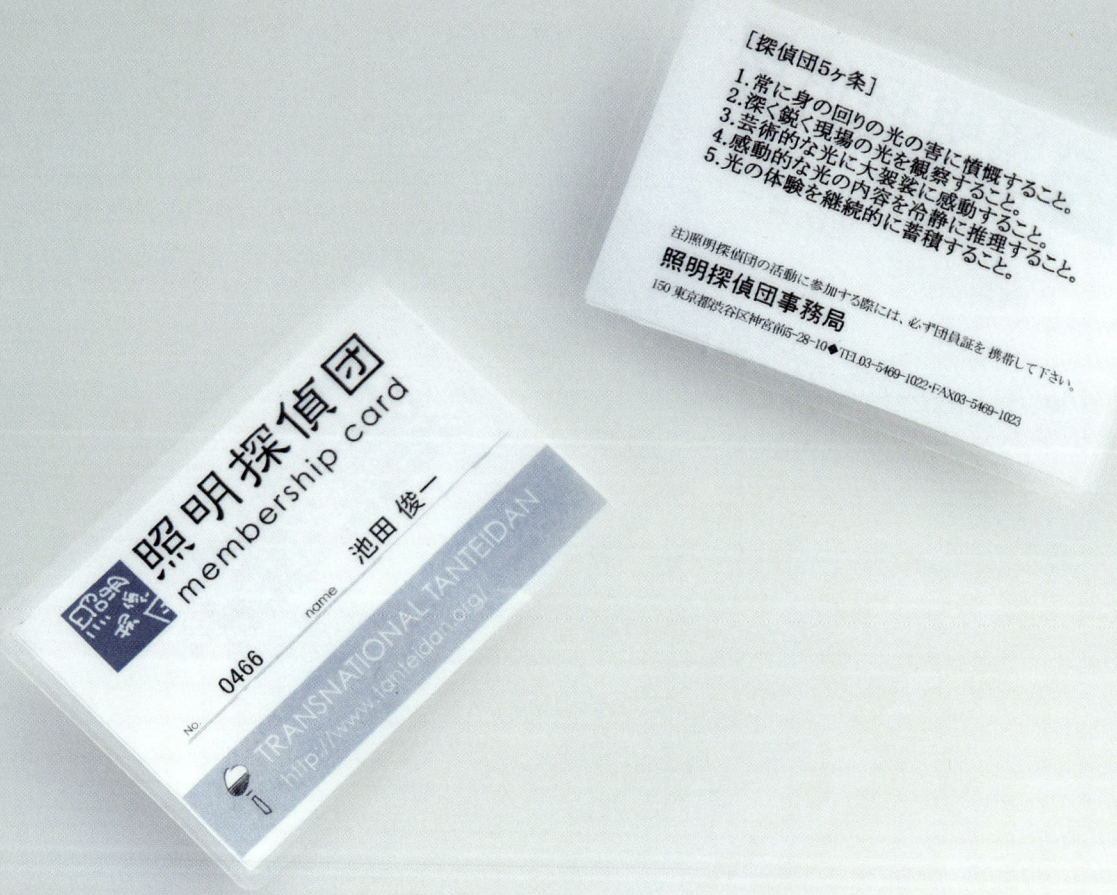

What is the Transnational Lighting Detectives aiming for?
For a better lighting world,
the Transnational Lighting detectives aim:
*Vision: To create an open platform to share inspiration
   and knowledge from local lighting cultures!
*Mission: To learn, respect and enjoy cross
   cultural lighting identities!
*Objectives: To talk and think about the future of light

Nature of Its Activities
The Lighting Detectives engages in mainly
seven activities as follows.
1 Daily city walks and salons
2 Study of urban lighting throughout the world
3 Publication and media
4 Seminars and exhibitions
5 Events in which the public is invited to participate
6 Communication through a newsletter and the Web
7 Transnational Lighting Detectives (TNT)

照明探偵団 7 つ道具
1 輝度計
2 トランシーバー
3 色彩照度計
4 距離計
5 ビデオカメラ
6 三脚
7 カメラ類
8 スケッチブック
9 地図
10 マグライト
11 身分証明書

The seven kinds of tools for Lighting Detectives
Luminance meter
Transceiver
Color temperature illuminance meter
Distance meter
Video camera
Tripod
Camera
Sketch book
Map
Maglite
ID card

光の世界旅行 ── ニューヨーク 2005　*World lighting tour in New York, 2005*

ニューヨークでのフォーラムはAIA（アメリカ建築家協会）センターで行われた。テーマは「メインストリート」ということで、世界各地の目抜き通りの照明環境が報告された。

　私たち東京のメインストリートは何といっても銀座通だ。銀座が日本の夜間照明を牽引してきたからだ。学生はニューヨークでの発表に持ち込むべき銀座八町の光環境をつぶさに調査した。1丁目から8丁目までは800メートルだが、その立体的な夜景を表現するために、道の両側から夜の立面を細かく撮影し、繋げたエレベーションを完成させた。プロジェクションのプレゼのみでなく、8メートルにプリントアウトした1/100画像を見せて喝采を浴びたことを思い出す。

The NYC Forum was held at the American Institute of Architects (AIA) Center. The theme was "Main Street" and lighting environments of main streets around the world were reviewed.

The main street of Tokyo is of course, Ginza Avenue because it has always driven lighting trends in Japan. Students surveyed the lighting environment of Ginza Avenue in detail from First Street to Eighth Street in preparation for the NYC presentation. To properly express the 3-dimensional scale of the street, students took pictures of and digitally recreated the elevation of the 800 meter stretched street. I remember, during the presentation students unrolled the eight-meter long printout at a scale of 1:100 of the elevation to a roomful of applause.

ニューヨークのフォーラムで紹介した
銀座の夜景エレベーション（部分）
*Elevation (in part) of the night view in Ginza, Tokyo
introduced in TNT Forum, New York*

AIAセンターでのフォーラム風景
吹き抜けた高天上が気持ちよい
*The TNT Forum at the AIA Center.
The vast atrium is excellent architecture*

ブルックリン・ブリッジを徒歩で渡った
イーストリバーから見るマンハッタンの夜景は格別だ
*Crossing the Brooklyn Bridge on foot.
The view of the Manhattan nightscape from the East River is one-of-a-kind*

前夜に調査した内容は1日で整理され、ボード展示された
Compiling information surveyed from the night before on a panel for display

短時間のプレゼ作業に表現技術が試される
As preparation time was very short, communication and expressive skills were put to the test

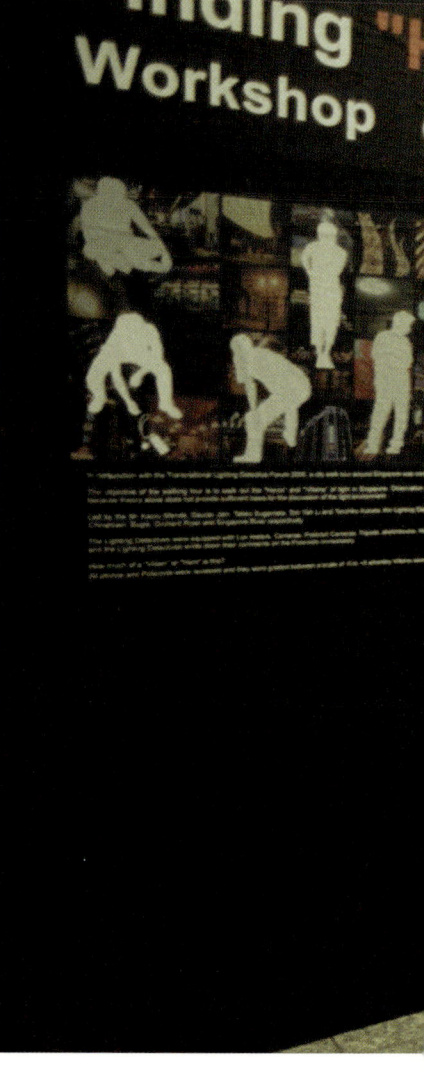

光の世界旅行 ── シンガポール 2006　*World lighting tour in Singapore, 2006*

シンガポールではTNTフォーラム始まって以来初めて、地元の人達とのワークショップが始まった。オーチャード、チャイナタウン、シンガポールリバー、公共団地、ブギス地区、の5つに分かれて学生と地元の人たちとの混成調査チームを結成した。

当初、言葉の不慣れな学生たちは戸惑いを隠せなかったが、現地を回り、調査パネルを作り、改善提案のスケッチをする段になると打ち解けた雰囲気で楽しんでいる様子だった。「光の英雄と犯罪者」というプラス・マイナスの評価表は大きなボードに張られて展示された。展示資料を基にプレゼンテーションとパネルディスカッションも行われた。

The TNT Forum in Singapore was the first and beginning of many workshops held with local participants. Divided into five mixed teams of students and locals, each team surveyed an area in the city including, Orchard, China Town, Singapore River, public housing complex, and Bugis.

At first, the unfamiliar language caused immediate confusion among the foreign students, but after completing the survey, panel work, and improvement proposal sketches together communication seemed to flow much easier in a relaxed and fun atmosphere. The large [Lighting Hero & Villain] evaluation panels rating lighting elements on a plus/minus scale were set for display and used as the bases for group presentations and follow-up panel discussion.

会場のURAセンターには照明探偵団の展示コーナーも作られた
A display corner at the URA center for the Lighting Detectives was also arranged

たくさんの聴衆を迎えて発表会とシンポジウムが開催された
In the face of a large audience, the presentation and symposium convened

光を観る
OBSERVING LIGHT

光の世界旅行
ベオグラード 2008
World lighting tour in Belgrade, 2008

セルビア共和国の首都ベオグラードはメジャーな観光都市とはいえない。それだからこそ濃密なワークショップが展開された。市街調査は、行政街、住宅街、商業地区という3種類に分かれ、例のごとく学生もバラバラに配属された。それぞれのグループには教育者を含む世界照明探偵団のコアメンバーが指導に当たり、夜遅くまで議論が続く。

次の日のワークショップの発表会は各グループごとにユニークなプレゼが行われた。その後のシンポジウムでは、聴衆の中に個性的な反論者が現れ険悪な雰囲気になる場面もあったが、その全てのアクシデントが貴重な経験となった。

上左｜照度を計る、ポラロイドカメラで撮影する、スケッチを描く……
上右｜提案をまとめる段になると声の大きい人の意見が通ることもしばしば
Upper left : Recording lux levels, snapping Polaroids, and drawing sketched were just some of the survey methods
Upper right : During discussion time, the louder voices were the more prominent opinions

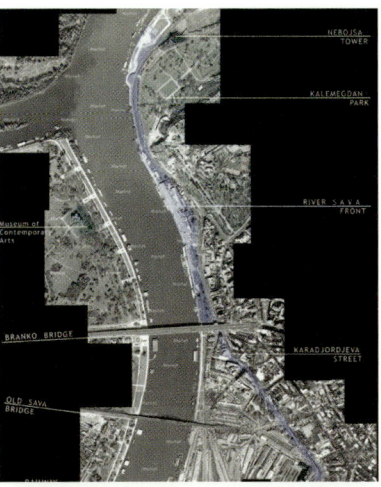

The capital of Serbia, Belgrade is not a major tourist city and this is exactly why we were able to hold a workshop of dense content here. The city survey was broken up into three areas, government, residential, and commercial and students were divided and assigned an area. An educator was also added to each group and under the guidance of the Transnational Lighting Detectives core members, surveys and discussions last long into the night.

The next day, each group gave a unique presentation at the workshop finale. At the following symposium, the atmosphere deteriorated amidst a strong-willed rebuttal from the audience, however all mishaps are now a valuable experience for TNT.

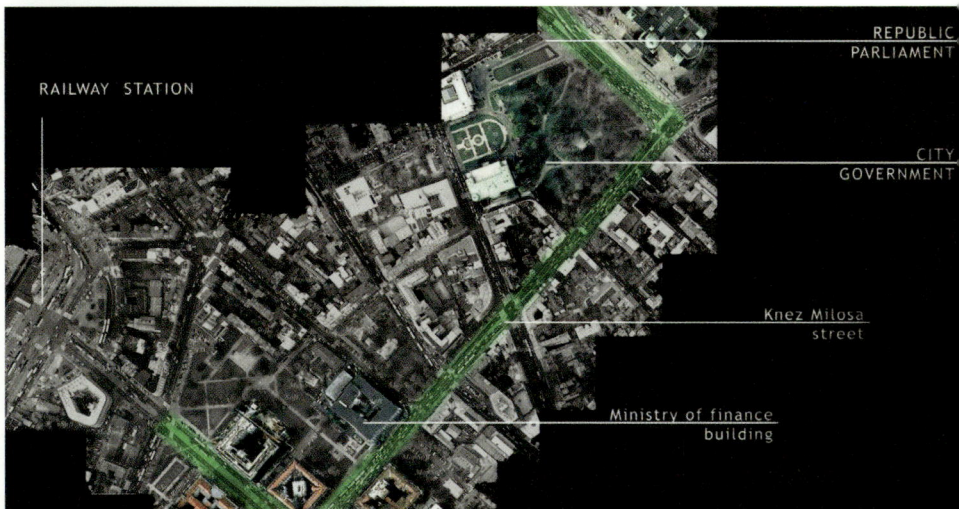

調査対象となった2種類の市街地
それぞれに2時間ほどかけて
徒歩による光環境調査が行われた
Pictures from the two areas surveyed.
Each team did about two hours of footwork
to address the lighting environment in each area

ベオグラードのTNTフォーラムで作成した
光の英雄と犯罪者のパネル例
＋の数が増えるのは英雄指数の高いもので
－は犯罪者として糾弾されるもの

*An example of the lighting hero and villain panels
compiled at the TNT Forum in Belgrade.
Examples on the + scale are heroes
and those on the – scale are condemned to villains*

LUMINANCE LEVELS

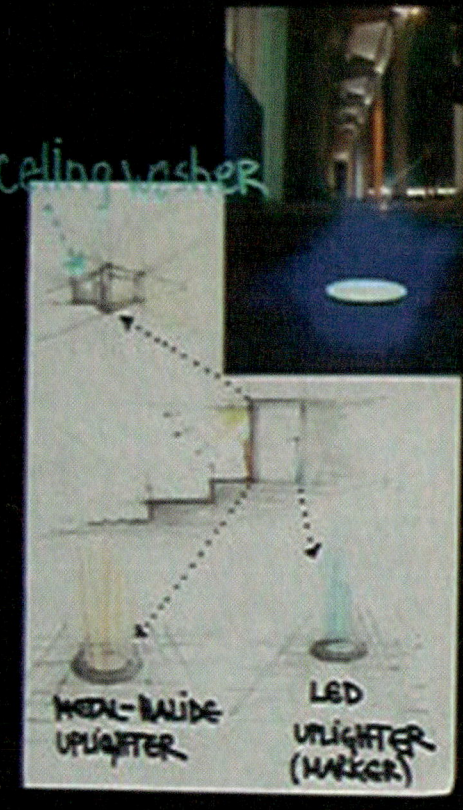

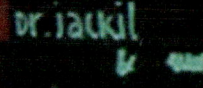

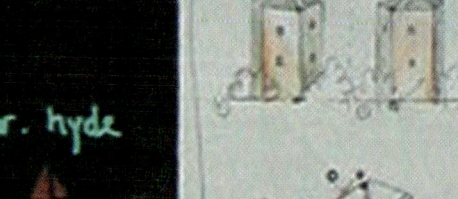

COLOUR PLAY

DEFINITELY MAYBE

+4

+3

+2

LIGHT PLAYGROUND

LIGHT GATE

+1

YOU CAN'T ILLUMINATE FAÇADE THROUGH THE TREES :)

−1

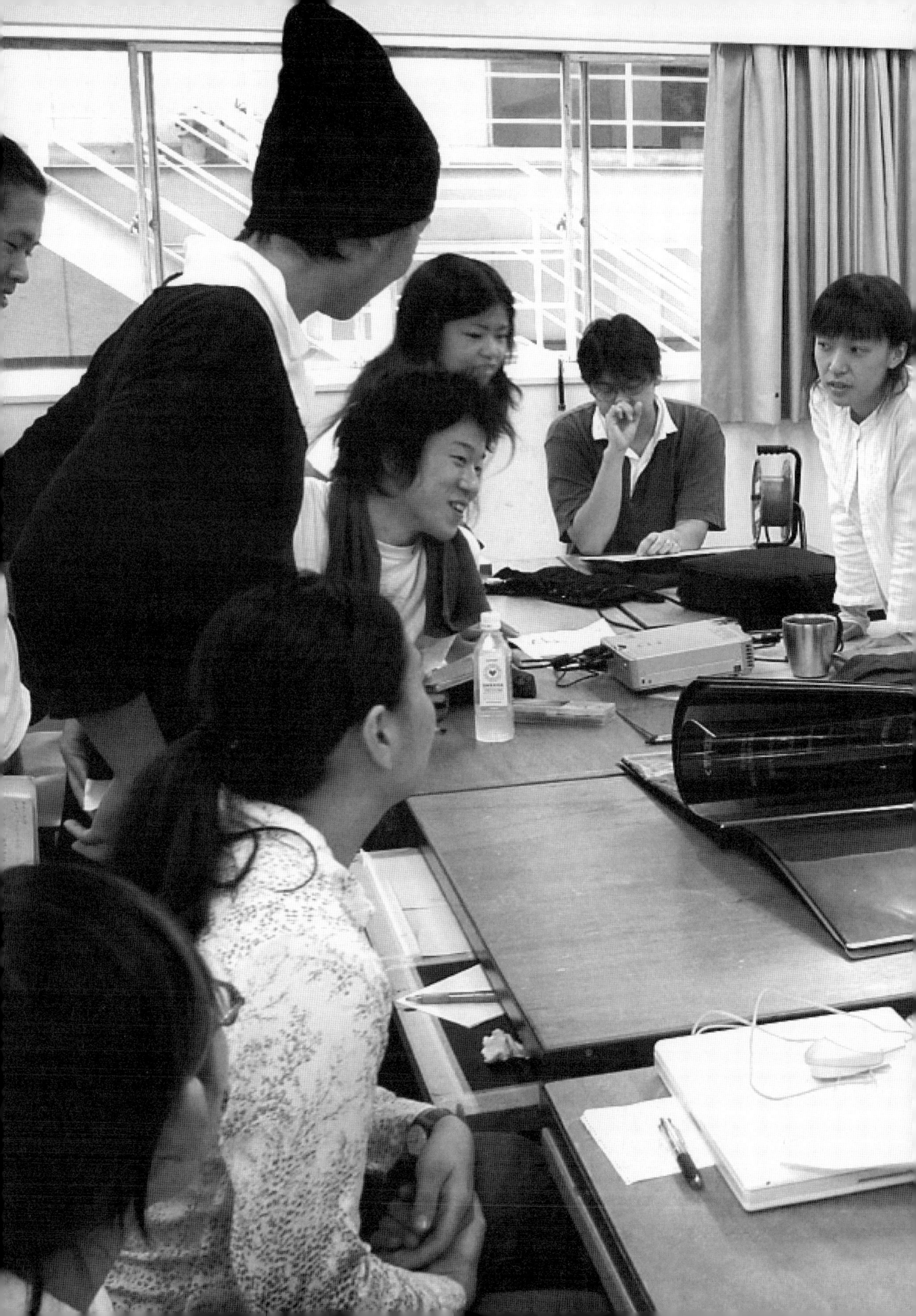

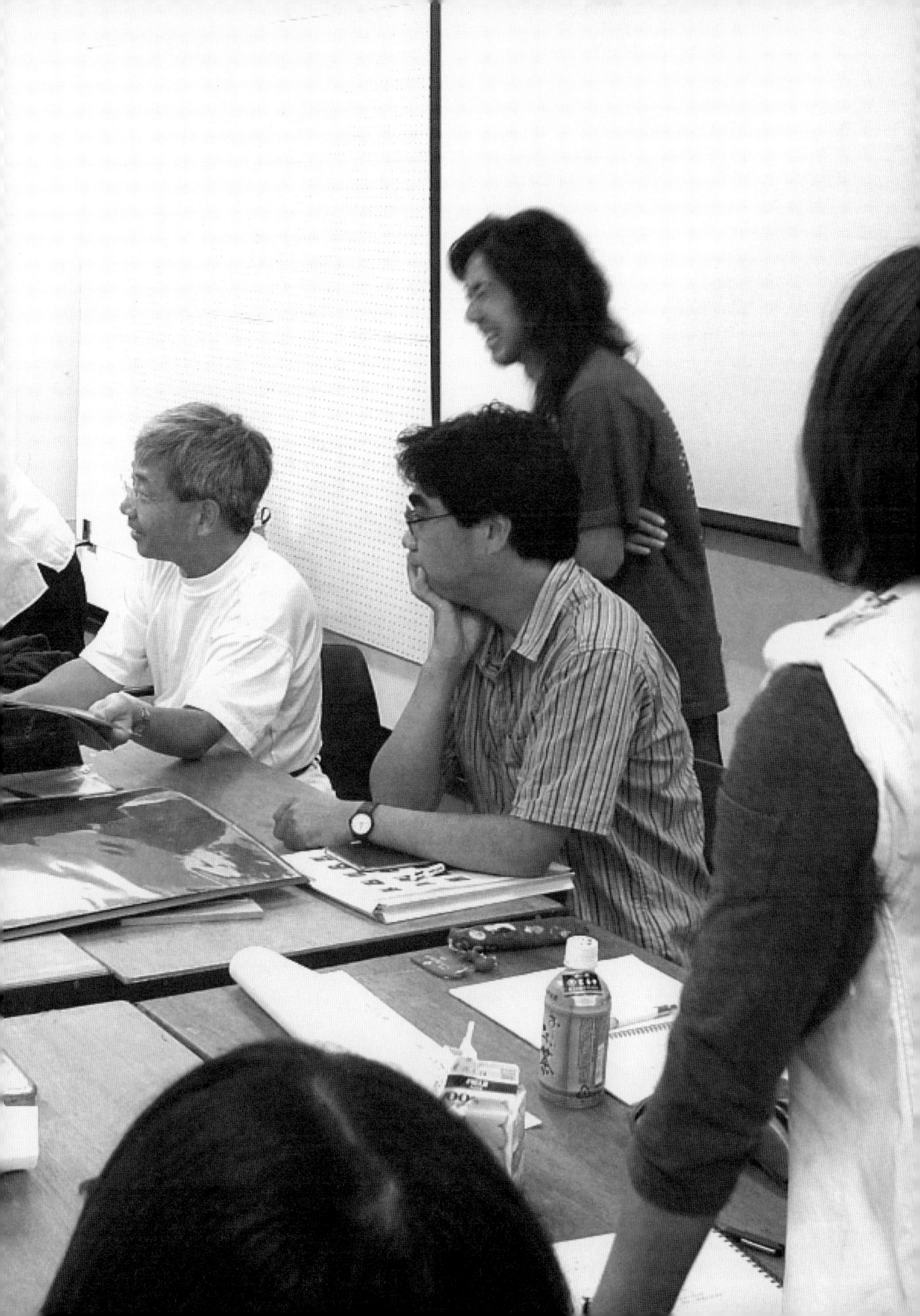

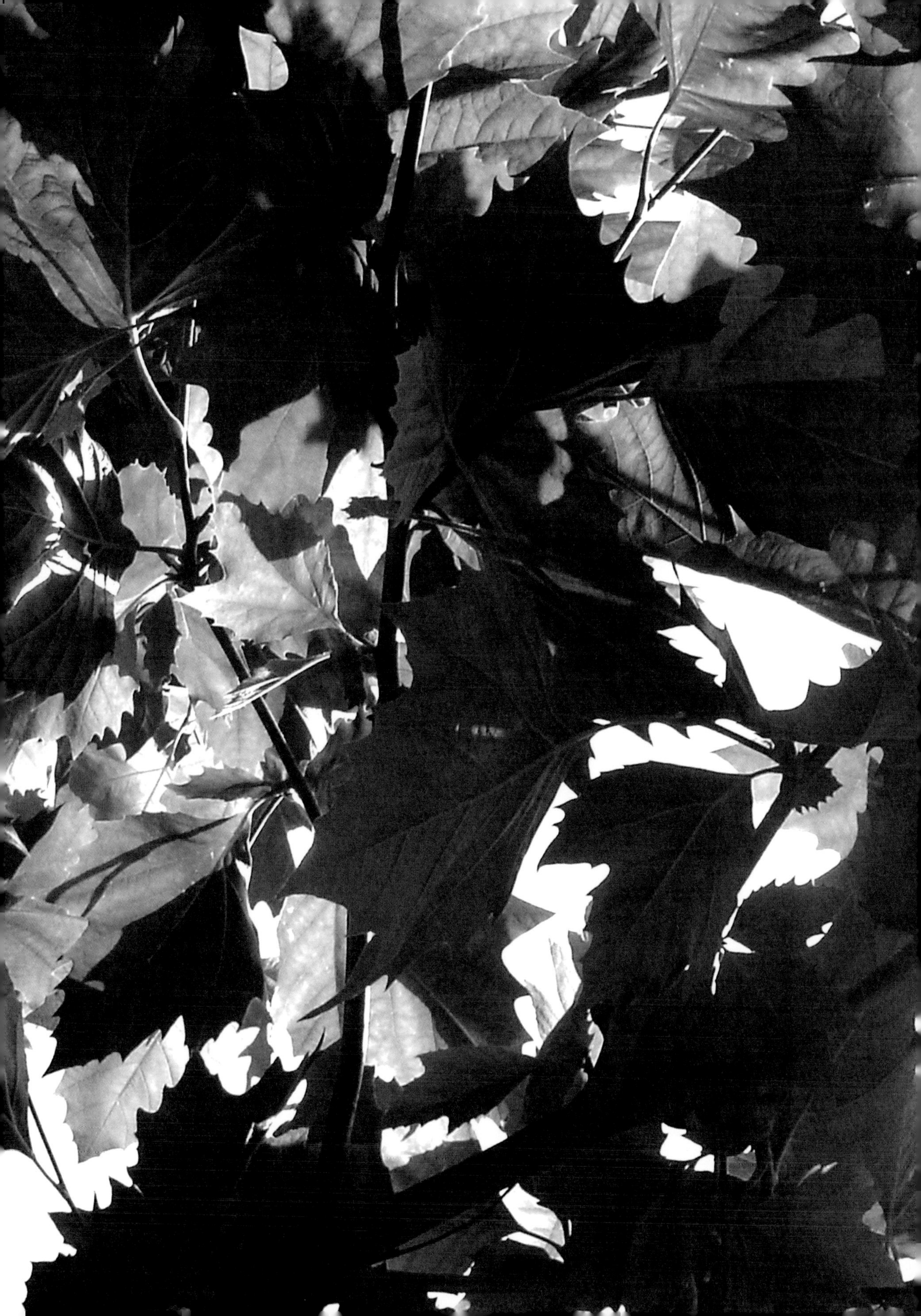

光を集める
COLLECTING LIGHT

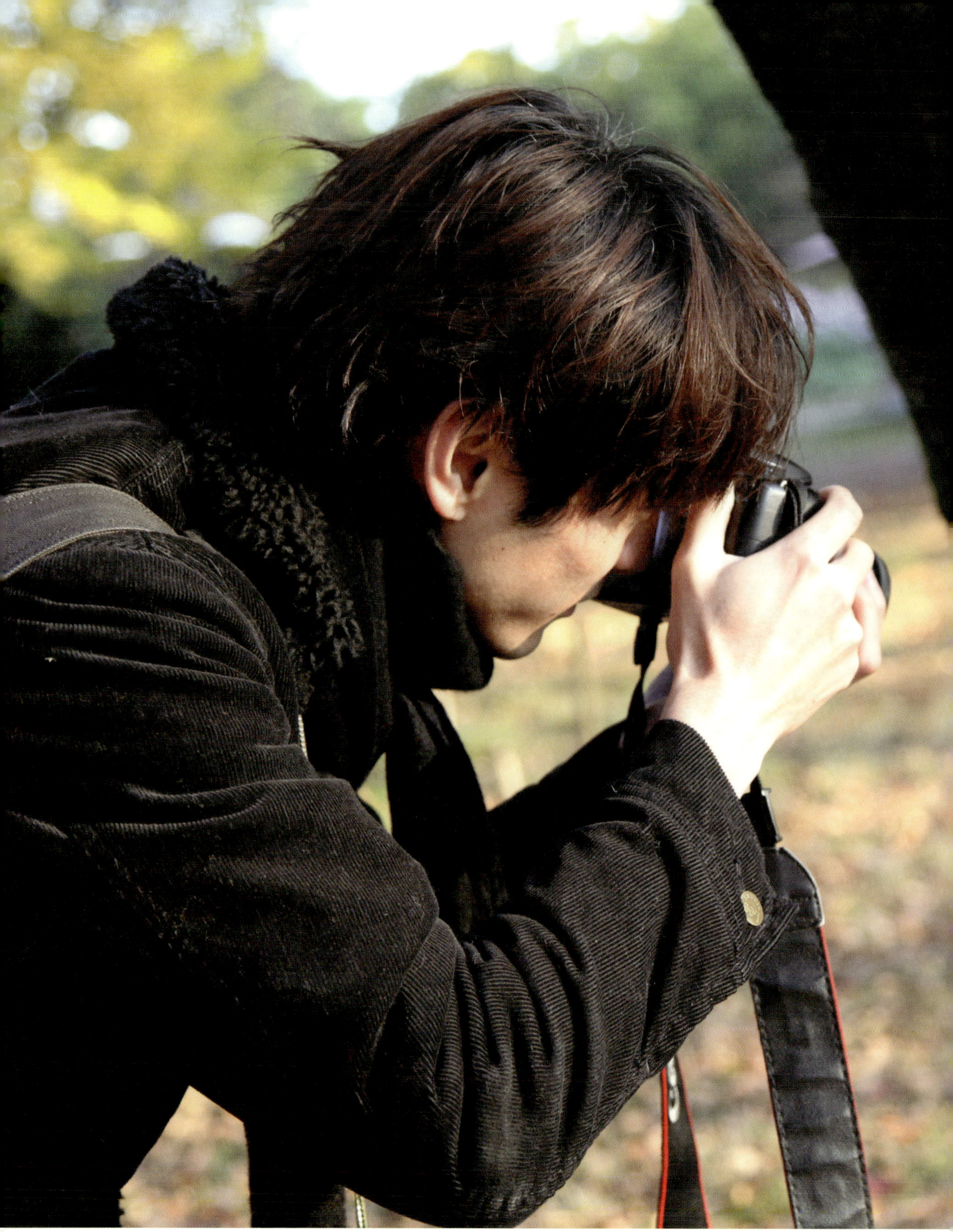

光を集める
Collecting Light

光を観るという基本動作を学習すると、次にそれを体に馴染ませるための学習に入る。それが光を集めるという行為だ。

　私たちは観光旅行に出るとたくさんの記念撮影をして旅の思い出を残そうとするが、光のデザインを学ぶ者は「観光」の意味を数段グレードアップしなければならない。様々な感動を与えられる場面において、視野内の景色から光をトリミングしたり、微細な光や影に限りなく接近したり、変化する光の状況を見極めたり、つまり昆虫採集ならず光採集を通して光の標本作りをするのだ。自然光を相手にした場合には快晴の日ばかりでなく、どんよりとした雲の下、あるいはしとしとと降りしきる雨の中、早朝の気配、昼に加熱する直射光、薄暮に染まる空、茜雲とブルーモーメント（透明な青い残照）、月あかりの海辺、満天の星降る夜、キャンドルの炎を囲む食卓、燃えさかるキャンプファイアーなど、それらのすべてに光採集のサンプルが潜む。大切なことは漫然と見とれずに、心惹かれる原因を冷静に探ることだ。正解をいい当てる必要はないが、何故その光を採集しようとしたかを自分の言葉でメモにする。それは時に詩情を含む散文の方が好ましい。現場での固有の感じ方を記録しておくべきなのだ。

　自然光相手でなく、陽の暮れた後の人工光の場合には光の採集の仕方が少し異なる。何故ならば、夜の帳を迎えて人工光が点灯する段になると俄に、私たちは慰められる光より、不快な光景を多く目にすることになるからだ。残念なことに都市照明の浅い歴史は、未だに快適な夜の都市景観を作っているとはいい難い。それ故に私は学生に対して「都市照明においては光の英雄と犯罪者を探せ」という注文をつけている。無条件に受け入れてよい自然光の場合とは異なり、夜間の都市景観に潜む光には批判精神が欠かせない。様々な光の事件に遭遇した時に、学生たちは躊躇なくその事件の張本人に△ではなく、○か×をつけることを強いられる。昨今の学生は「カワイイ〜」「おいしい〜」といった安易な英雄採点が多く、犯罪者を厳しく糾弾する勇気に欠ける傾向があるのが残念だ。この「自然光と都市光を採集する」という学習方法は、ライトコレクション（Light collection）と命名され、ゼミの中心的課題として欠くことなく継続して出題された。

Having learned what to do first—that is, to see light—one begins to accustom oneself physically to that act by collecting light.

When we go on a sightseeing tour, we take many photographs to serve as mementos of the trip. For those who are studying the design of light, "sightseeing" must mean a great deal more than casually visiting places of interest. They need to collect "specimens" of light as entomologists collect insect specimens; when they come across scenes that move or impress, they must isolate light from the landscape, view from close up details of light and shadow, and ascertain the changing conditions of light. In the case of natural light, the weather is not always good. There are many different specimens of light to be collected: an overcast sky, gentle incessant rain, the first light in the morning, the hot, direct sunlight of noon, the sky at twilight, clouds made crimson at dawn or dusk, the blue, transparent vestige of light in the sky known as "the blue moment," a moonlit seashore, a night sky filled with stars, a dining table lit by candles, a blazing campfire. It is important to coolly analyze what makes a light attractive; one must not simply look on aimlessly. Accurately describing everything is not necessary, but one needs to make a note in one's own words of why one is interested in collecting a particular specimen of light. At times, prose that captures a poetic sentiment may be preferable to facts. One ought to record perceptions unique to the scene.

Collecting specimens of artificial light after the sun has set requires a slightly different approach. That is because at night, when artificial lights have come on, one is more likely to encounter light that is unpleasant rather than comforting. Unfortunately, urban lighting does not have a long history and cannot be said to have produced many pleasant urban nightcaps. For that reason, I tell students that, in the case of urban lighting, they must search for either "heroes" or "villains" of light. Whereas natural light can be accepted unconditionally, a critical spirit is indispensable to the observation of light in urban nightcaps. When they come across an exceptional circumstance, that is, an "incident" involving light, students must not hesitate to label those responsible as either heroes or villains. Unfortunately, students today are too easy on wrongdoers, approving anything that seems to them cute or stylish. They tend to lack the courage to deal severely with criminals. The collection of specimens of natural light and urban light—what I call "Light collection"—is an important task that I have always assigned to students in the seminar.

ライトコレクション
Light collection

———

デイライト・コレクション

* 雲・霧・水、緑・木・石・土・ガラス・砂、朝日・夕日・建築・障子・窓などなど…。自然光を受けているものの表情、かたち、素材、テクスチャー、光沢感、色、湿度や温度まで、光を注意深く観察する。
* 身の回りの自然光を深く観察し、その興味深い表情（情景や状態）を100種類採集しなさい。デジタルカメラを使用する。
* 次にその画像の中から25種を選択しなさい。それぞれの画像を140ミリ×140ミリ正方形サイズにトリミングし、B1サイズ内に5×5=25に配置しなさい。
* 25種の画像の中から最終的にベスト3を選出し、それぞれに100字以内の詩をつけなさい。
* 提出物の内容は、25コレクションのB1パネルとそれを収めたデジタルCD。ベスト3はA3サイズ3枚に画像、タイトル、詩をコラージュする。
* プレゼンテーションはベスト3を、詩の朗読付きでプロジェクションする。

アーバンライト・コレクション

* 夜間になると都市は数多の光に満たされてくる。商業の光、オフィスから溢れる光、住宅の窓から漏れる光、色とりどりに輝く広告サインなど…。ドキドキさせられる光や、うっとりする光もあるが、イライラさせられる光も少なくない。それらを注意深く観察する。
* 私たちの身の回りにある人工光を注意深く観察し、1光の英雄、2光の犯罪者、3先端的な光、を自由に採集しなさい。
* 興味を持った都市光を100種類採集し、その中から25種類を抽出する。自分にとっての「英雄光、犯罪光、先端光」に分類しなさい（分類の割合や枚数は自由）。
* 作業や提出物の内容、そしてプレゼ方法などは上記の自然光と同様。
* プレゼンテーションは1＋2の両者で一人6分以内とする。

(i) Daylight collection

* Observe light carefully in terms of the expression, shape, material, texture, glaze, color, humidity and temperature of such objects as clouds, fogs, water, greenery, trees, stones, soil, glass, sand, rising sun, setting sun, architecture, shoji or paper sliding screens and windows that receive natural light.
* Carefully observe natural light around yourself and collect 100 examples of interesting expressions (scenes and conditions) using a digital camera.
* Select 25 out of the images you obtained. Trim each image to a square 140 mm wide and 140 mm long and arrange 25 (5 x 5) images in a B1-size panel.
* Select best three out of the 25 images and write a poem for each in less than 100 words.
* Submit a B1-size panel carrying a collection of 25 images and a digital CD containing the images. For the best three images, develop three A3-size sheets carrying the image, title and poem.
* Project the best three images and recite the poems in presentation.

(ii) Urbanlight collection

* Cities are filled with numerous types of light at night such as the light of commercial activities, light overflowing offices, light pouring out of the windows of houses and multicolor light of advertisement boards. Some types of light thrill or enchant you but not a few irritate you. Observe such types of light carefully.
* Observe artificial light around yourself carefully, and freely collect (i) heroes, (ii) villains and (iii) vanguards among light examples.
* Collect 100 urban light examples that intrigued you and select 25 thereof. Classify the selected examples as heroes, villains or vanguards from your viewpoint. (You may determine the percentages and quantities of classifications at your discretion.)
* The work, materials to be submitted and presentation methods are the same as for natural light collection described above.
* Time assigned to presentation for (i) and (ii) above shall not exceed six minutes per presenter.

デイライト・コレクション
Daylight collection

見慣れている光も改めて観察してみる
A fresh look at everyday light

心に留まった光をカメラにおさめる
Point the camera at light that grabs you

アーバンライト・コレクション
Urbanlight collection

照度計を用いて数値を計る
Carry a luminance meter to record lux values

時には怪しいホテル街を散策することも
Sometimes we even walk suspicious streets with dodgy hotels

	A	B	C	D	E
1					
2					
3					
4					
5					
6					

44

デイライト・コレクション
5つの選択画像の評
Daylight collection
Analysis of 5 selected images

1-D
海に沈む夕日/ほんの10分間に昼と夜が交差して、オレンジ色の光は優しく激しく変化する。

2-G
ペットボトルの水がレンズになる。焦点を得た光と拡散してにじむ光が交差する。

4-A
柔らかい葉の表面にゆったりと陽があたり、そこに影と陰が現れた。

5-G
誰もが心奪われる木漏れ日/光ではなく竹林に揺れる影にこそ、私たちは魅了されている。

6-C
海面を反射する太陽光をグリッタリングと呼ぶ。これだけはギラギラではあるが不快感がない。

1-D
A sunset sinking into the ocean/For 10 minutes everyday, night and day intersect and the warm, but intense orange light transforms.
2-G
Water in a plastic bottle becomes a lens. The light from the focal point and diffused and blurred light intersect.
4-A
Sunlight pools on the soft surface of a leaf, creating light and shadow.
5-G
Mesmerizing sunbeams filter through the foliage./It's not the light, but because the shadows are in a bamboo grove they are captivating.
6-C
Sunlight reflecting off the surface of the ocean is referred to as a glitter ring. Even thought the light is glaring, it is by no means unpleasant.

甲斐蓉子

Yoko Kai

世界中を旅する甲斐は幅の広い自然光に出会ったはずだ。自然光は時に寂しく、時に暖かく、時に不可思議に見えたに違いない。光と影の強い対比に惹かれた感情の起伏が手に取るようだ。

Kai is a world traveler and has had the opportunity to experience a wide range of natural light. Natural light has many faces; sometimes desolate, sometimes inviting, and sometimes amazingly mystical. This work clearly shows the strong contrast between light and shadow that attracts rugged emotion.

廣瀬文音
Ayane Hirose

廣瀬は感じ取る能力が高い学生で、人に見えないものも見えているかのようだった。それだから時々、彼女との会話は難解だった。切り取られた自然光はどれも柔らかくぼんやりするものが多い。

Hirose's capability to sense things is exceptionally high as a student. She seems to see things that others can't or won't and as her professor, our conversations were very abstract. Many angles of natural light in her work seem soft and hazy.

伊藤一実

Hitomi Ito

この25種類の自然光はどれもじっとしていない。揺れていたり、流れたり、時間をかけて移動していたり…。自然光が見せる表情には個性的な息づかいがあることを、伊藤が気付かせてくれている。

All 25 types of natural light in this work are constantly on the move; wavering, flowing, transforming over time, etc. The various expressions that natural light shows are individual and seem to be breathing. Ito's work is a great reminder of this.

刈谷康時

Yasutoki Kariya

それぞれの自然光に、きりっとした緊張感がある。漠然と情景を見ていない凛々しい姿勢を感じる。刈谷は誰にでも優しい男だが見る目に険しさがある。感動の深さがそうさせるのだろうか。

All the examples of natural light in this work show a clear sense of tension. The posture of each is distinguished, and not vaguely looked at from behind the viewfinder. Kariya is a young man who is kind to all, but has a very sharp eye. I believe this sharp eye is a product of the depth of his emotion.

	A	B	C	D	E
1					
2					
3					
4					
5					
6					

アーバンライト・コレクション
5つの選択画像の評
Urbanlight collection
Analysis of 5 selected images

5-B
浅草雷門の大提灯。この前で多くの観光客が記念写真を撮影する。どうして提灯を外から照らすのか？

2-D, 3-F, 4-G, 5-G
繁華街のネオンサインを学生たちは嫌うが、私はネオン街の中にも良し悪しを見分ける。

4-E
有楽町ガード下の飲み屋街は赤提灯の良さを忘れていない。しかも天井間接照明つき。

5-E
大観覧車が水面に映えている。これを英雄として称える人も犯罪者と蔑む人もいる。

5-H
鉄骨工事現場で激しく火花を飛ばす職人さん。夜間の工事現場はライトアート・ギャラリーのように興奮する。

5-B
The giant lantern at Asakusa Kaminarimon is a popular spot for tourist to take a souvenir photograph. Why do they insist on illuminating the lantern from the outside?

2-D, 3-F, 4-G, 5-G
The students find the neon lights of the entertainment districts distasteful, but I think you can to discriminate the good from the bad.

4-E
The drinking establishments under the railway at Yurakucho have not forgotten the charm of red lanterns and their ambient light.

5-E
A giant ferris wheel reflects off the water. Some people would praise this as a lighting hero, but others despise this nightscape is a villain.

5-H
Sparks fly at a steel frame work construction site. At night the construction site seems to transforms into a light art gallery.

奥田啓晃

Hiroaki Okuda

切り取られた夜の景色の幾つに奥田は心を許したのだろうか。
この25枚の中には光の英雄と犯罪者が潜んでいる。私が犯罪者
のラベルを貼るものには、奥田は英雄のラベルを貼るに違いない。

How many of the nightscapes seen here did Okuda really approve of deep down in his heart? The 25 images are a mixture of light heroes and villains. I bet, those that I label as villain, Okuda would label hero.

興松麻美
Asami Okimatsu

興松は好き嫌いをはっきりいえる学生だった。直観に頼って良いから潔く○×をつけて状況を評価することが大切だ。この25枚の都市夜景の断面には彼女の叫びと強い詩情が感じられる。

Okimatsu is a student who can clearly state her likes and dislikes. Being able to rely on intuition and clearly say "yes" or "no" is an important skill. I can sense her cry and strong poetry in her 25 images of urban nightscape.

光に詩を添える
Writing a poem for light

ライトコレクションは単に光を採集するにとどまらない。その光や影がどんな状況で環境に潜んでいて、どのようにして自分の眼や心を捉えたのかを記録しておく必要がある。私たちはデジカメでシャッターを押すとともに、思い思いのメモ帳や情報カードに自分の言葉を書き記すことを常とする。

しかし、光や影の感動を写真でしか伝達しえないことを補うために、私は学生に一振りの詩を強要する。デザイナーにとって言葉は大切な武器だと思うからだ。1枚の画像に込めた深い気持ちを学生は言葉で補い、それを自ら音読することで、表現者としての資質を学習する。

Light collection is not simply the collection of light examples. It is necessary to record under what conditions of the light and shadow are hidden in the environment and how they are caught in your eyes or mind. We routinely press the shutter button of a digital camera and write down words on a notepad or information card that we like.

I make it mandatory that students write a poem to supplement the communication of the thrills that they obtain from light and shadow examples in addition to using photographs. This is because I believe that words are an important tool for designers. Students expand on their deep feeling that they put into an image, using words, and master a qualification for expressing themselves by reciting the poem by themselves.

明ける　興松麻美

渋滞もたまにはいいなと思う
群青と赤の色がなんだかとても清々しい。
苛々しがちなテールランプも
赤信号もいい仕事をしてくれるときもある
そろそろ日が出る頃
赤信号よまだ動いてくれるな。

透明　齋藤麻里

ところどころ濁った
色とりどりの宇宙みたいに
星が見える
薄くとがった
とけていく体温みたいに
花が咲く
まろやかに眠って
鮮やかな夜が明ける
もう姿のない結晶

常磐線　北村康恵

よぎる。
着崩した制服
鞄からはみだした参考書
置き去りにされた日本酒のカップ
陸橋を超える　おと　ひかり
発車ベルの瞬間
すぎる。
わたしが変わっても
みんなが変わっても

あまぐも　　刈谷康時

カレにしたら
不猟で
体毛は湿気て重く
頭まで痛み始めた

雨ナンテ キライ
いいえ　雨は贈り物
シャンデリアを愛で
一句練り
甘露をいっぱい
雨ハ 命ノ 恵ミ
うるおう心
推し量れぬクモ心

ジューシー絨毯　　刈谷康時

びっしり集まり
すべすべ滑らか
ジューシー絨毯
緑の海原
裸足で踊る
がっちり蔓延り
たっぷり水っ腹
ジューシー絨毯
翠の深海
静かに眠る
異分子排除
苔国家
乾燥注意
ジューシー絨毯
団結のわたつみ

今
裸でコロコロ
塩分被弾

透明　齋藤麻里

ところどころ濁った
色とりどりの宇宙みたいに
星が見える
薄くとがった
とけていく体温みたいに
花が咲く
まろやかに眠って
鮮やかな夜が明ける
もう姿のない結晶

朝食　吉田尚加

冬の朝6時　目が覚める
ふとんの中でぐずぐず
カーテンの隙間から
優しい光が射し込む
スイッチが切り換り
私はキッチンへ
パンはトースターへ
卵は油をひいた熱々のフライパンへ
冬の朝7時　窓際の朝食
太陽はぐずぐず昇る途中

面出ゼミの誕生とその意味　杉本貴志

私が空デに主任教授として在席していた間の大事な仕事の一つが、必要な時に専任教授候補を選定し、教授会に推薦する事であった。会社内の昇進人事とは異なり、選定される新しい教授はその能力、人となりだけではなく、この科のあるべき将来像や、当然教えを受け卒業してゆく学生等の将来に深くかかわる事になるだけに、その人の現状だけではなく専門とされる領域そのものの可能性に一歩踏み込んで考察されねばならないし、随分重い仕事であった。

面出先生について考えた時は、私が主任になった最初の人事案件であった。まず考えたのは空デという領域についてで、武蔵美には工芸・建築という科が存在し、そのうちでインテリアデザインは以前から開講されていた。空デは、私が着任した頃から学生の志望も強くなりデザイン系としてそのあたりを俯瞰し、なんとか教育が行われるようになっていたが、所謂インテリアデザインとは少し趣が異なっていたように思っている。

インテリアデザインという領域は、本来人が生活する事がそのベースで、そこからスタートをするべきであるのだが、我々の社会は縦横無尽に発展を遂げ、社会の骨格が大きく変化をしている。当然、生活というシンプルな一言では表せなくなっている。

食べ物を見ても我々の口に入るものは畑や海や川、山や牧場から運ばれるのではなく、スーパーやデパート、コンビニや駅の周辺の店舗から購入しているし、どこかのメーカーによって製作されパックされているものがほとんどで、まずそれらを時間をかけて吟味しなくてはならない。衣類といえば大量の情報として我々に与えられ、選択するためには膨大な知識が必要である。TVのコマーシャルや雑誌、カタログを抜きには語る事はできない。我々が住む住宅や家具も同様である。我々の社会は、ほとんどがそんな情報によって成り立っているといわざるを得ない。

ある意味では虚構である。そこにデザインの多くが存在している。我々の歴史のうちでの50年、100年というのは、流れる時間のほんの一部であるのだろうが、その短時間を辿ってもその大きな変化に驚かざるを得ない。しかし、その事に併行して変わらず今に繋がるものも見逃すわけにはいかない。

例えば、

　「願わくは　花の下にて春死なむ
　　　その如月の望月の頃（西行）」

という歌を聞くと、800年以上過ぎてなお、言葉の香りのまま我々の頭に入ってくるし、その情景が美しく目に浮かぶのである。気持ちよく同調するのである。――こういう万葉集以来の無数の言葉、風習、絵画、風景に繋がって、我々は形をなし、あるいは作られてきたし、その最後尾で悩み、新しく作ろうとしているのである。空デの専任には、これらの両性が必要なのだと私は考えた。一つは、現代社会と正面から対峙し、その混沌に一刀を振るう知力を持ち、一方に流れる豊饒な香りに心を震わせながら悩むべきである。

面出先生は、新内の名手でもある。彼に着任について相談をした時、まずきっぱりと辞退をされた。理由は仕事が忙しいとの事で、まず当然であろう。10回位、昼夜を問わず数ヶ月に渡って説明をした。そうした上で、長文の手紙を書いた。口で喋る事と文字で読む事は少し違うのである。彼が昔独立して会社を作る時、少々私が手伝った事も忘れずに書き添え、脅迫紛いな事まで書いた。私は、空デに彼が欲しかったのである。余人ではすまなかったのである。その結果、諦めるように着任し、面出ゼミが発足した。

ある意味では、学生たちは幸せであったのだろう。一番大きかったのは、面出先生を通じて学生も時代と向き合ったのである。面出先生、永い間、ご苦労様でした。

杉本貴志　*Takashi Sugimoto*

インテリアデザイナー/1945年東京都生まれ。1973年株式会社スーパーポテト設立。「無印良品」の名店舗はじめ、商業空間のデザインを数多く手掛ける。海外のホテル、バー、レストランの内装デザインから複合施設の環境計画、総合プロデュースまで活動は多岐に渡る。武蔵野美術大学名誉教授。1985年から2010年まで「TOTOギャラリー・間」運営委員。株式会社春秋代表取締役。

Interior designer/Born in Tokyo in 1945. Founded Super Potato in 1973. Worked in numerous commercial space design projects including the shops of Ryohin Keikaku Co., Ltd. Activities range widely from the interior design of hotels, bars and restaurants in other countries to environmental planning and comprehensive production of composite facilities. Professor Emeritus, Musashino Art University. TOTO Gallery-Ma steering committee member in 1985 through 2010. Representative Director, Shunju Co., Ltd.

The Birth and Significance of the Mende Seminar

One of my important jobs when I was chairman of the Department of Scenography, Display and Fashion Design was to select and to recommend to the faculty a candidate for full-time professor when the position became open. The job was a heavy responsibility in that it required careful consideration of not just the person's present circumstances but the potential of that person's field of specialty because, unlike a matter of promotion within a company, it was more than just a question of ability and character; the selection would have great impact on the future image of the department and the future of the students who are taught and will graduate from the university.

Deliberating the appointment of Professor Mende was the first personnel matter I dealt with after I became departmental chairman. The first thing I considered was the field of scenography, display and fashion. Musashino Art University already had Departments of Architecture and Industrial, Interior and Craft Design, and classes were already being held in interior design. Students had begun to express a strong interest in scenography, display and fashion around the time I took up my position; examining the design-related curriculums offered at the time, I saw that it was possible to get an education in that area of study. However, it seemed to me slightly different in character from interior design.

Interior design is premised on people actually using the spaces that are designed in their everyday lives, but our society has developed in many different directions and is undergoing major changes in structure. "Everyday life" is no longer a simple concept.

The items of food we put in our mouths are not transported from seas, rivers, mountains or pastures but purchased from supermarkets, department stores, convenience stores and stores near railway stations. Most of it is prepared and packaged by some maker, and we have to take the time to select from the items available. As for clothing, we are provided with an enormous amount of information and require a great deal of knowledge to make our selections. No discussion of the issue would be complete without a consideration of television commercials, magazines and catalogs. Much the same can be said of the houses we live in and the furniture we use. Our society for the most part is dependent on such information.

In a sense, that aspect of society is invented. Much of design exists in that world of fiction. Half a century or a century is only a a brief interlude in the flow of time of history, but the great changes that can take place even in that short time is amazing. However, we cannot overlook the things that remain unchanged and connect the past to the present.

For example, when we read the poem by Saigyo, *"If I could have my wish, I would die in spring under cherry blossoms around the full moon in the second month,"* the words still resonate in us and the image they evoke is still vivid even though more than 800 years have passed since they were written. We identify with the poet. Connected to the past by countless phrases, customs, paintings and landscapes, we have shaped and created forms, and now, at the tail end of this long chain of connections, we worry over and attempt to create new things. I felt that a full-time professor of the Department of Scenography, Display and Fashion had to be aware of both these aspects. He would have to have the intellect to confront contemporary society head-on and cut through its confusion but also be sensitive to our connections to the past.

Professor Mende is also accomplished at Shinnaibushi, a style of narrative music using the shamisen. When I first asked him to accept the position, he declined outright. He said he was too busy with his work, which was what I had expected him to say. I talked to him about ten times, at all hours of the day, during a period of several months and then wrote him a long letter. A written argument is slightly different from a spoken one. I did not neglect to add that I had helped him in the past when he was opening his own office and resorted to a bit of emotional blackmail. I wanted him in the Department of Scenography, Display and Fashion. No one else would do. In the end, he gave in and accepted the position; the Mende seminar began.

The students, in a sense, were fortunate. The most important thing was that the students, through Professor Mende, have been able to confront the age they live in. Thank you, Professor Mende, for your many years of effort.

光に触れる
Touching Light

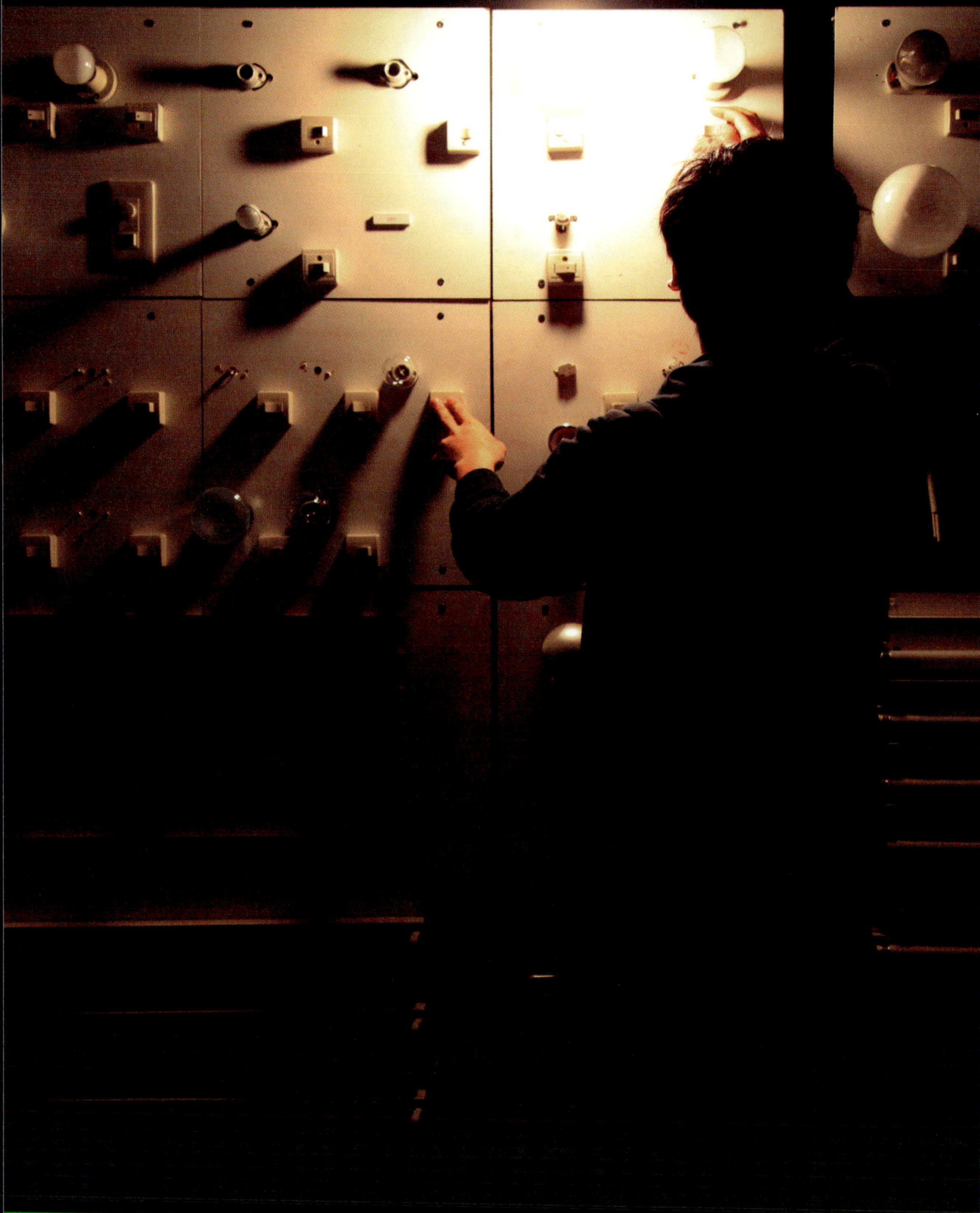

光に触れる
Touching Light

現場に出て先入観なしに光に出逢い、その光をよく観察しそれを最終資料化するという行為が基本だと解説したが、次に光のデザインは多様な光の実態に触れてみて初めて、光の本質に迫ることができる。光は嘘をつかない。そして光は思うようにいうことを聞いてくれないことが多い。光に触れて火傷を起こすこともあるし、光ではなしに電気に触れて感電することさえある。多種の光に触れあい、光とのもめ事を重ねて、早く光と深い仲になってしまう必要がある。

自然光の特性を知るために、太陽の直射光を様々なフィルターやレンズで加工実験をすることもある。また、1本のローソクの炎によってできる揺らめく陰影や、燃えさかる囲炉裏の火を正確に制御するのも技術の鍛錬だ。しかし、何といってもこの100年に満たない短い期間に人類が開発してきた数多の人工光源から発する光の種類と品質は、現代照明デザインの基礎となる。この電気エネルギーを用いた各種光源の発する光の特性を知り、更にそれらの光源を用いて光のデザインを展開するために考案された各種の光学的照明器具の性能を知ることが、光を学ぶ者の基礎的作業となる。

光のゼミナールではまず、各種の光源から発せられる光を間近で観察できるように、点灯できる光源のサンプル台(ライトボックス)を作成した。低圧ナトリウムランプ、水銀ランプ、メタルハライドランプなどの高輝度放電灯から、蛍光ランプ、冷陰極管、ハロゲン、クリプトン、キセノンなどを用いた各種白熱ランプ類、そして現代の英雄とされるLEDまで。それらの光源を一堂に並べてみると、1世紀ほどの現代照明デザインが俯瞰される思いだ。

光学的な照明器具についていうと、光源からの光を反射鏡(リフレクター)で配光制御する形式のもの、様々な光学レンズ系で制御するもの、そしてフードやバンドアといった遮光技術を応用したものなどを用いて、意図した光の形をオブジェクトに与えていく作業になる。

光源種の違い、使用電力や光の出力の違い、そして光の色(色温度)や色の再現性(演色性)、器具から飛び出てくる光の形(配光)の相違など。実際に重たい照明器具を現場で扱ってみて初めて分かることばかりなのだ。その理解のレベルを高めていくことを置いて光のデザインは学べない。

As I have explained, to study the design of light, one must go out into the field, encounter light without preconceptions, carefully observe that light and transform what one has observed into usable data. The next step is to come into contact with the diverse actual conditions of light; here for the first time, one pursues the essence of light. Light does not lie. Moreover, light often does not do what one tells it to do. One can burn oneself touching light or, even get a shock if one touches an electrical source. One needs to come into contact with diverse forms of light and enter into a close relationship — a relationship that may include friction or discord — with light as quickly as possible.

To become familiar with the characteristics of natural light, one also needs to experiment with direct sunlight using diverse filters and lenses. Learning how to precisely control the flickering shadows produced by a single candle or a fire blazing in the hearth is also part of the training. However, the types and qualities of light generated by the many man-made sources of light that humankind has developed within the last one hundred years are basic to contemporary lighting design. Learning the characteristics of light generated by different light sources using electrical energy and the capabilities of the different types of optical lighting fixtures that have been devised for the purposes of lighting design using those different light sources is a basic task for anyone studying light.

In the Light Seminar, we first prepared sample light boxes that could be switched on so that light generated by different light sources could be observed closely. High-intensity discharge lamps (such as low-pressure sodium lamps, mercury-vapor lamps and metal-halide lamps), fluorescent lamps, cold-cathode fluorescent lamps, and incandescent lamps (using elements such as halogen, krypton and xenon), and today's much-touted LEDs — seeing the lights from all these light sources arranged in a single space is like getting a bird's eye view of a century of contemporary lighting design.

With respect to optical lighting fixtures, the task is to throw light onto objects in the intended form through the use of devices that control light distribution by means of reflectors or optical lenses and devices such as hoods and barn doors that shield light.

The differences in light sources, the differences in the amount of electrical power used and in the output of light, the differences in color (or color temperature) or color reproduction (or color rendering) capability, and differences in the shape of light (or light distribution) generated by the fixture — these are things that can only be understood when one has actually dealt with heavy lighting fixtures on the scene. One cannot learn the design of light without continually raising the level of that understanding.

ライトアップゲリラ　*Light-up guerrilla*

アノニマスな都市の景観を光で再生する
Reproducing an anonymous urban landscape using light

グループ・4週間　　　　　　　　　　*Assignment for groups / Four weeks*

この課題はキャンパスを離れ街に出て行う光の操作演習である。各種の光源や照明器具を使い、アノニマスな都市の景色を創造的に創りかえることを目的としている。各種の光に直接触れ、操作することを通じて、光の即興的性質を学習する。

先ず始めに大学キャンパス周辺を夜間調査し、光の再生を必要としている現場を探し出しなさい。特徴ある景観が闇に埋もれてしまっている環境、明るすぎて折角の環境が光によって破壊されているところ、ちょっとした光の演出を施せばきめんに価値を増大させるような場所、などをロケハンして情報を持ち寄る。
次にその情報の中から3種類の現場を選択し、ライトアップの手法を検討し実験計画書を作成する。
当日は2台の車に必要機材を搭載し、ライトアップ実験計画書に従って手際よく作業を行い、各種の記録をとる。
短時間の街を相手にした実験なので役割を分担し、手際よく成果に結びつける。
事前と事後の状況との解りやすい比較資料が大切になるので、まとめ方を決定してから事件の手順や内容を考えるようにする。
ライトアップゲリラ報告書を作り、その成果をゼミ発表や空デHPにアップする。

The assignment requires that students go out of campus and manipulate light in town. The objective is to creatively reproduce an anonymous urban landscape using various light sources and lighting equipment. Students learn the extemporaneousness of light as they come into direct contact with light and manipulate light.

-First, make night surveys around the university campus to find sites where light needs to be reproduced. Go location hunting to gather information on sites where a unique landscape is buried in darkness, where excessive light spoils a good environment and where the value would be immediately improved with slight dramatization of light.
-Select three sites, examine light-up methods and develop plans for experiment.
-On the day of implementation, carry necessary equipment in two vehicles, implement the light-up plan efficiently and keep records.
-Share responsibility and produce results efficiently because the experiment needs to be conducted in a short time in town.
-Fix the methods of preparing materials for comparing the conditions before and after implementation prior to the planning of the procedure and details because such materials are important.
-Prepare a Light-up guerrilla report and present the results in seminar presentation or upload them on the website of the department of Space Design, Musashino Art University, .

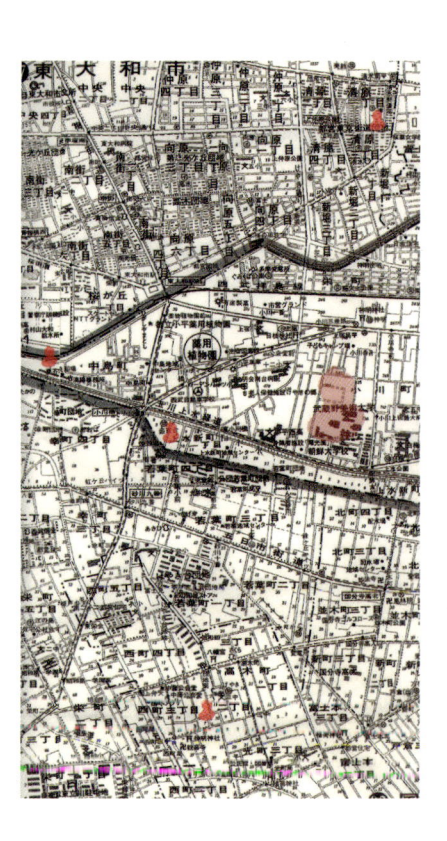

ライトアップゲリラの実行場所（キャンパス周辺地区）
Map of Light up guerrilla executed (the surrounding area of the campus)

使用機材
Machinery and tools for use

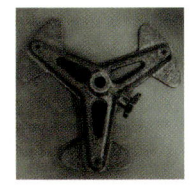

カットスポット	レフランプ（100w）	パーライト（ナロー）	調光機	三脚
Cut spot	*Reflector lamp*	*PAR light*	*Dimmer*	*Tripod*
クリップソケット	ベース	延長コード	ミニレフソケット	ゼラケース
Clip socket	*Base stand*	*Electric cord*	*Mini reflector lamp socket*	*Filter case*
移動用台車	ドラム	ガソリン	紐	発電機
Carrier	*Cable drum*	*Gus tank*	*Rope*	*Generator*

日程		Schedule	
11/16	課題のオリエンテーション （ゲリラ候補地の選定、役割の分担）	11/16	Guerrilla orientation (Select potential locations, divide duties.)
11/17-29	自主的なロケハン作業と計画案の エスキース作成	11/17-29	Scouting for locations and sketching proposals.
11/30	面出中間チェック （ゲリラ計画書の作成、作戦会議）	11/30	Midterm check with Prof. Mende. (Written guerrilla proposal and team meeting.)
12/05	ゲリラ実習本番（照明実験）	12/05	Light-up guerrilla Practical Event (Lighting experiment)
12/07	講評（成果発表とゲリラ報告書の作成）	12/07	Review (Report results and prepare written report.)

Light-up guerrilla 1

ゴミ捨て場
Garbage dump

小平市のゴミ焼却施設を学生が探し当ててきた。通常は高く積まれたゴミ捨て場などに誰も目などくれるものではない。しかし夜の帳が訪れると闇に紛れてゴミの山は視覚的に姿を消失するに違いない。そして1つ1つ丁寧に解釈を与えた光を重ねて行けるとしたら、もしかすると昼に汚いゴミの山は、夜に芸術的なオブジェとして蘇るかもしれない。それが私たちの狙いであり野心であった。

Students found a waste incinerating plant in the city of Kodaira. Nobody generally takes notice of a pile of garbage. When night falls, the pile of garbage may fade away into darkness. The pile of garbage, dirty during the day, may be reborn as an artistic objet at night if rays of light are applied in layers to each of which careful interpretation is given. So we expected and we were driven by an ambition.

Before

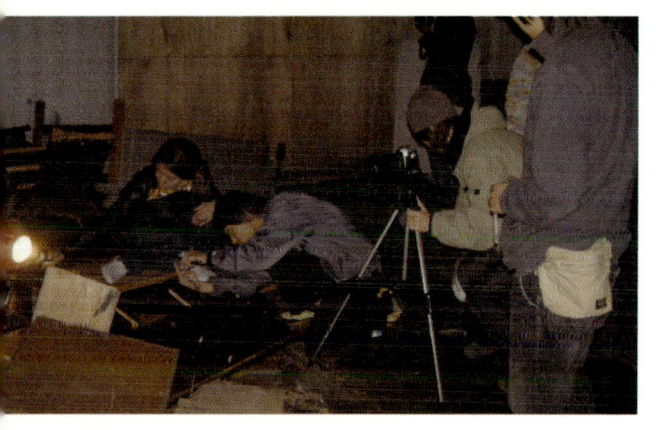

デジカメ
指揮官
面出先生
林先生
カメラマン

After

光に触れる
TOUCHING LIGHT

67

Light-up guerrilla 2

民家
Old private house

昔は日本酒か醤油を作っていたような民家だが今は使われていない様子だった。何よりも周囲に街灯が少なく闇の中に沈んだ表情と背景の大きな樹木が魅力的だ。大きな面積の外壁には板が貼られているので、白熱灯のスポットライトで柔らかく照らし上げるだけで、暖かい民家の香りがしてくるようだった。

　無許可の仕事は手際よさが求められる。もちろん通りすがりの見物人などを作ってもいけない。さっと照らしてさっと逃げる。これがライトアップゲリラの鉄則だ。

The old private house seemed to be uninhabited although it may have been used to produce Japanese sake or soy source. What made it attractive were the atmosphere as it was sunk in darkness with no street lights in the vicinity and the tall trees in the background. Boards were on large exterior walls. Simply illuminating the boards softly with incandescent spotlights was expected to cause the house to emit a warm scent.

Unauthorized work required deftness. There had to be no onlookers at all. Just light up quickly and run away. That is a rigid rule of Light-up guerrilla.

Before

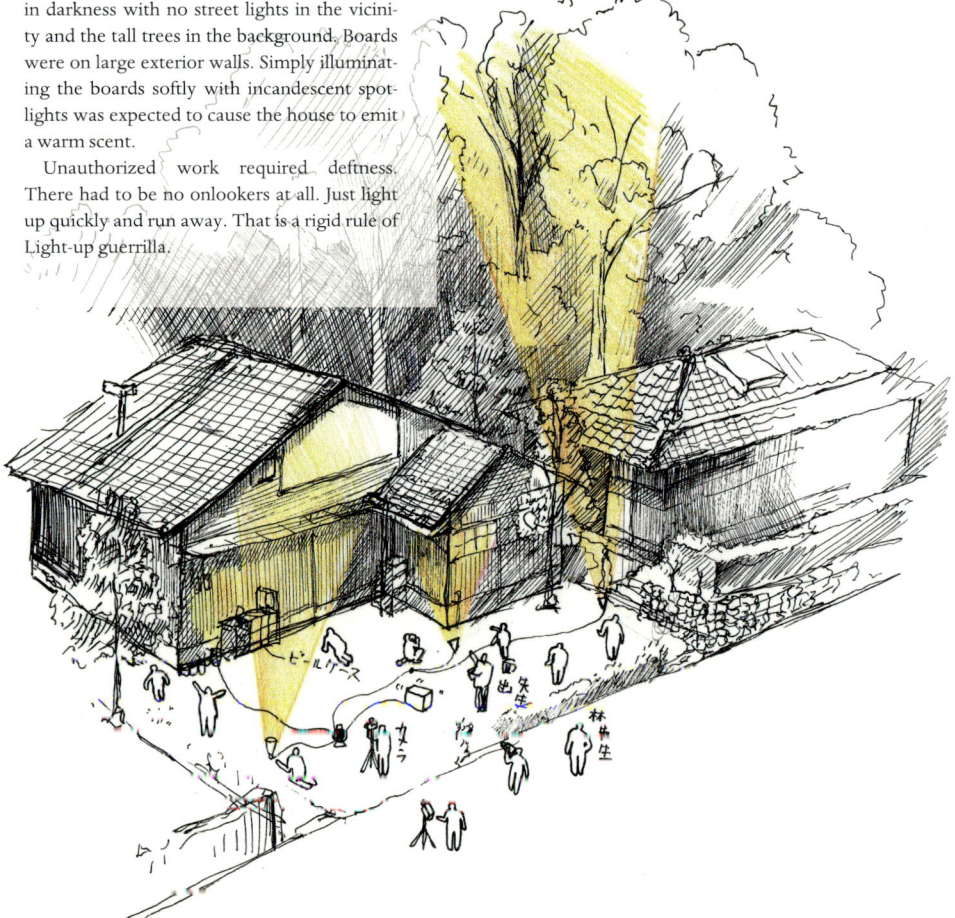

After

光に触れる
TOUCHING LIGHT

After

街灯は光をあてることで消灯する
街灯を消して、ゲリラの準備
Shine direct light on the streetlights to turn them off.
Now the guerilla lighting can begin

Before

Light-up guerrilla 3
児童公園
Children's park

ジャングルジムと砂場があるだけの児童公園に面して、今はほとんど入居者のいない2階建ての公団住宅が建っていた。住宅の陸屋根に旧式なテレビアンテナが寂しそうに並ぶ姿も学生たちの心を捉えていた。昼に風化した景色だが夜に光を当てると過去の記憶までもが蘇るような景色に一変した。

ジャングルジムの線材を冷たく輝かせる光は、そのままに住宅棟の外壁に入り組んだ格子の影を投影した。砂場を斜めに照らす光と影は、昼間に遊んでいた子供たちの時間を記録している。何とアノニマスな街角は芸術的なのであろう。

A two-storied apartment building of the Housing and Urban Renaissance Agency with few inhabitants was located opposite a children's park only with a jungle gym and a sandbox. The old-fashioned lonely TV antennas on the deck roof intrigued students. Lighting the weathered landscape as seen during the daytime turned into a landscape that was likely to revive past memories during the night.

The light that coolly illuminated the linear members of the jungle gym cast the shadows of intertwined grids on the exterior wall of the house. The light that diagonally illuminated the sandbox and its shadow were the records of children who played in the daytime. How artistic an anonymous street corner could become!

Light-up guerrilla 4
ビニールハウス
Plastic greenhouse

東京都小平市といっても大学の周辺には小規模農家を営む人も少なくない。学生たちは農業用のビニールハウスを発見した。ハウスは内部の高揚感だけでなく、中でうごめく光と影を外部から鑑賞する楽しさもある。彼らは各種の色光を試し、自前のバイクのヘルメットに無数のアクリミラーを張り付けた。人間ミラーボールは農家の一角をディスコに早変わりさせた。もちろん、このビニールハウスをお貸しいただいたお宅には、菓子折りと記念写真をもって丁寧にお礼に伺った。

Not a few households engage in small-scale agriculture around our university located in the city of Kodaira, Tokyo. Students found an agricultural plastic greenhouse.

Plastic greenhouses not only bring a sense of exaltation when you are inside but also provide you with joy of watching the light and shadow moving inside from out of the greenhouse. Students tried light of varying colors and attached innumerable acrylic mirrors to their own motorcycle helmet. The human mirror balls turned part of the farm house into a disco. Of course we called on the owner of the greenhouse with a box of cake to express our sincere thanks.

ミラーボール・ヘルメット
Special helmet like mirror ball

上｜ミラーボールによって拡散された光
左｜ミラーボール・ヘルメットを装着
右｜ビニールハウス外観

U: Diffused light by the mirror ball
L: A man equipped the helmet of the mirror ball
R: Appearance of plastic green house

光に触れる
TOUCHING LIGHT

ノートパソコンを用いプロジェクタで照射
Projecting lighting images via a laptop computer

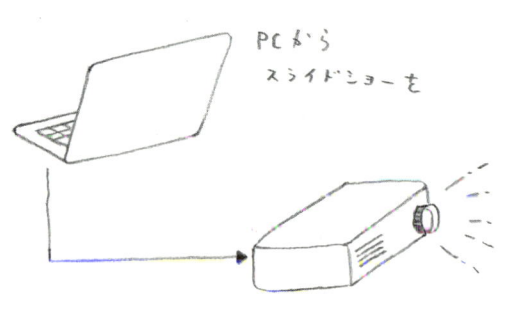

PCから
スライドショーを

プロジェクタで
投影する

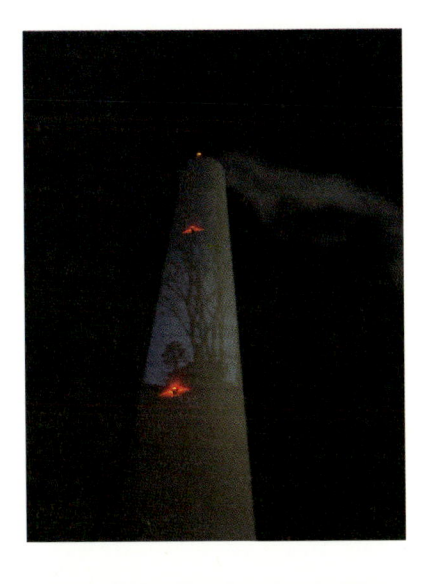

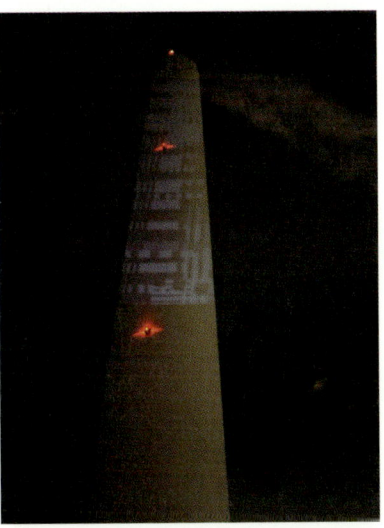

Light-up guerrilla 5

煙突
Chimney

直径5メートルもありそうな巨大な煙突がそびえている。地域のゴミ焼却施設に付帯するものだ。そこから立ち上る煙と遥か上空を流れる白い雲がこのオブジェに不思議な芸術的価値を与えていた。このような産業構造物が地域の芸術的アイコンに変容するかもしれない…。学生たちは「嫌われている環境をアートする」という課題に取り組んだ。

A five-meter-diameter giant chimney stood attached to a local waste incineration plant. The rising smoke and white clouds flowing above gave the objet a mysterious artistic value. The industrial structure might turn into a local artistic icon. Students worked on an assignment, "making a hated environment into art".

ライトボックスの制作
Making of Light box

「自然光に学べ」というのがゼミ語録のようになっているが、光に触れるという学習は太陽光や灯火と戯れることのみを意味しない。むしろ難解なのは、この100年余りに人類が開発してきた無数の人工光源の性質を正しく識別することである。

低圧ナトリウムと高圧ナトリウムランプ、水銀ランプ、メタルハライドランプ、直管型・環型・U字型・ボール型の蛍光ランプ、シリカ型・クリア型の普通ランプ、クリプトンランプ、ハロゲンランプ、低電圧ランプ、そして各種の発光ダイオード（LED）など…。

光源の開発は日進月歩だ。光に触れることの基本は人工光源を自ら点灯してその性質を知ることである。学生たちはそれぞれ単独に点滅できる光源装置（ライトボックス）を製作した。

"Learn from natural light" is something like a message of the Mende Seminar. The study for touching light does not simply mean playing with sunlight or lamps. More difficult is to accurately understand the properties of innumerable artificial light sources that humankind has developed in the past 100 years. Artificial light sources include low- and high-pressure sodium vapor lamps, mercury lamps, metal-halide lamps, straight circular, U-shaped and ball-shaped fluorescent lamps, silica and clear lamps, krypton lamps, halogen lamps, low-voltage lamps and various types of light emission diode lamps. Light sources are being developed continually. The basic objective of touching light is to understand the property of an artificial light source by turning it on by oneself. Students made a light source device (Light box) that they can make blink independently.

分類	No.	名称	電力(W)	備考
白熱電球 IL	1	クリアランプ	60	
	2	ホワイトランプ	60	
	3	シルバボウル電球	60	
	4	ハンドミラー電球	60	
	5	レフランプ	100	屋内用
	6	レフランプ	100	屋外用
	7	ミニレフランプ	40	
	8	ミニクリプトン	60	クリア
	9	ミニクリプトン	60	ホワイト
	10	シャンデリア電球	40	クリア
	11	シャンデリア電球	40	ホワイト
	12	ビームランプ	60	集光形
	13	ビームランプ	60	散光形
ハロゲン HAL	14	ミニハロゲン	50	12V
	15	ミニハロゲン	100	110V
	16	ハロゲン	50	12V・N
	17	ハロゲン	50	12V・M
	18	ハロゲン	50	12V・W
高輝度放電灯 HID	19	蛍光水銀灯	100	
	20	透明水銀灯	100	
	21	リフレクタ形水銀灯	100	
	22	高圧ナトリウム灯		NH70F
	23	高圧ナトリウム灯		K-HICA150BF・G
	24	低圧ナトリウム灯	90	
	25	メタルハイドランプ	70	CDM-TD
	26	高演色ナトリウムランプ	100	
	27	メタルハイドランプ	70	CDM-R
	28	メタルハイドランプ	70	CDM-TT
	29	メタルハイドランプ	70	CDM-T
蛍光灯 FL	30	直管Hf蛍光灯16形	16	P電球色
	31	直管Hf蛍光灯16形	16	P温白色
	32	直管Hf蛍光灯16形	16	P白色
	33	直管Hf蛍光灯16形	16	P色
	34	直管Hf蛍光灯16形	16	Pday色
	35	直管T5	14	
	36	ブラックライト蛍光灯	20	
	37	コンパクト蛍光灯	16	FHT16W形
	38	コンパクト蛍光灯	18	FML18W形
	39	コンパクト蛍光灯	18	FDT18W形
	40	コンパクト蛍光灯	18	FPT18W形
LED	41	調光調色型LED電球		DL-L60AV リモコン付き
	42	一般電球型		LDA9L-H
	43	ボール型LED電球		DL-L81A
	44	ミニクリプトン型LED電球		LDA5-E17
	45	小型電球型斜め取り付け専用		LDA6D-E17/BH
	46	シャンデリア電球型LEDフィラメント	0.6	ULEF100v-0.6w25/C3E17
	47	ミゼットレフ電球		LDR6L-W ミゼット
	48	ビームランプ型LED	5.3	LDR14L-W ビームランプ
	49	ハロゲン電球型LED電球	5.3	LDR7L-ME11
	50	ダイクロハロゲンLED	6.8	MASTER LED Lv FAN付き
	51	LEDラインユニット		2700K
	52	LEDラインユニット		3200K
	53	LEDラインユニット		3600K
	54	LEDラインユニット		4500K
	55	LEDラインユニット		5300K
	56	LEDラインユニット　カラータイプ		自動で色が変化
	57	直管蛍光灯形LED		R-FAC40BN1　P昼白色
	58	直管蛍光灯形LED		R-FAC40BD1　P昼光色
	59	直管蛍光灯形LED		R-FAC40BL1　P電球色

ライトボックス　立面図
Elevation of Light box

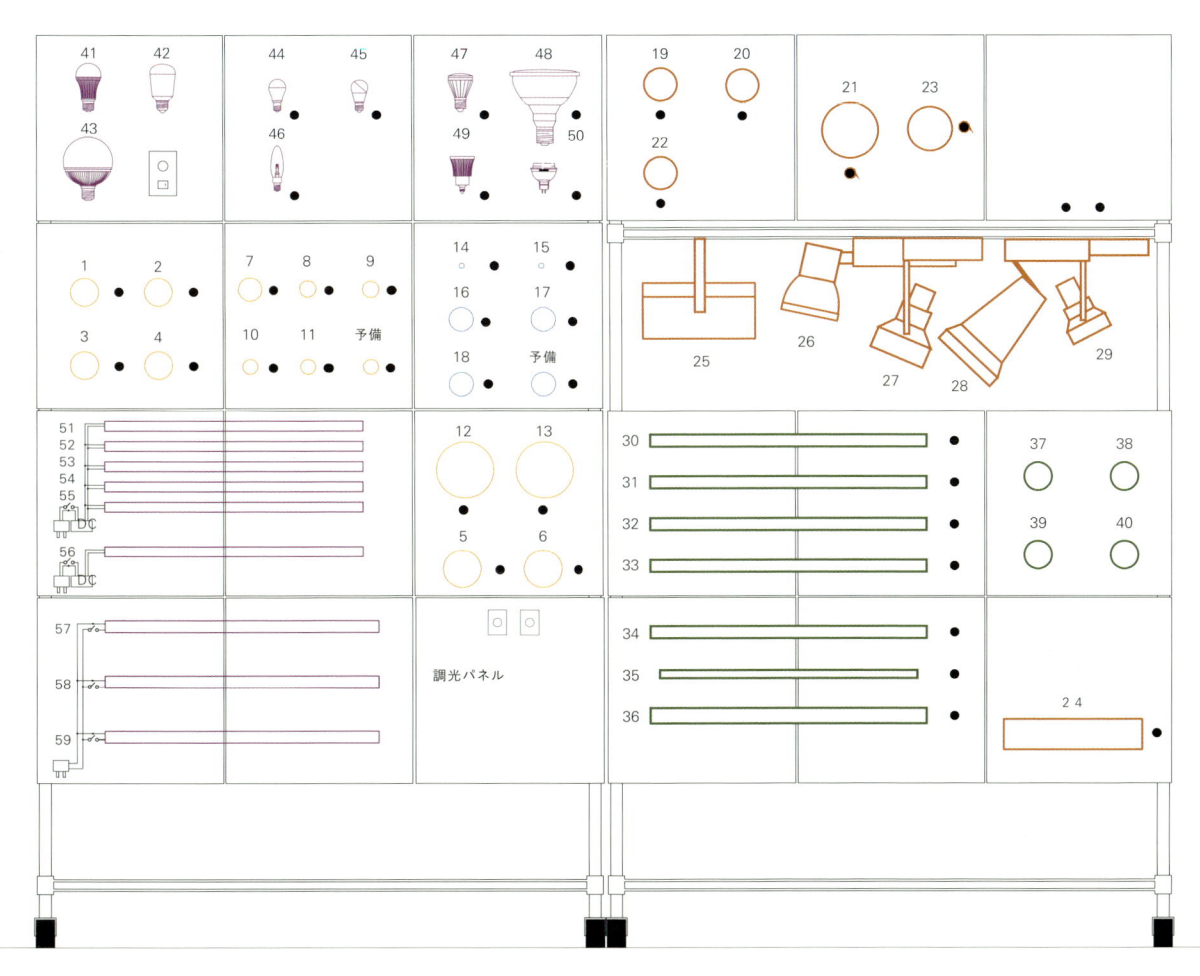

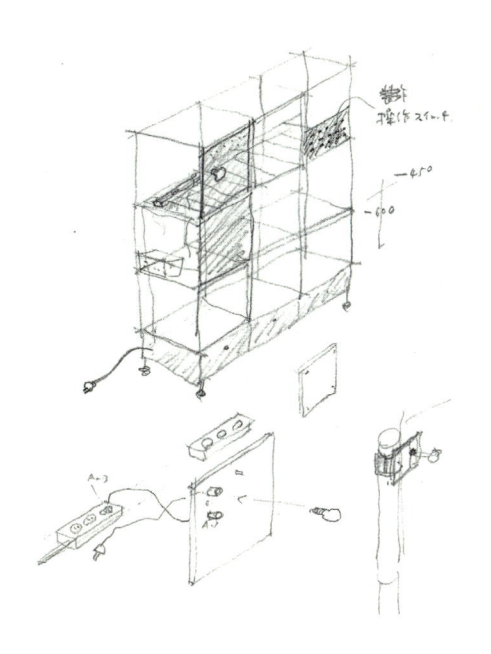

＊●印：点灯スイッチ

光に触れる
TOUCHING LIGHT

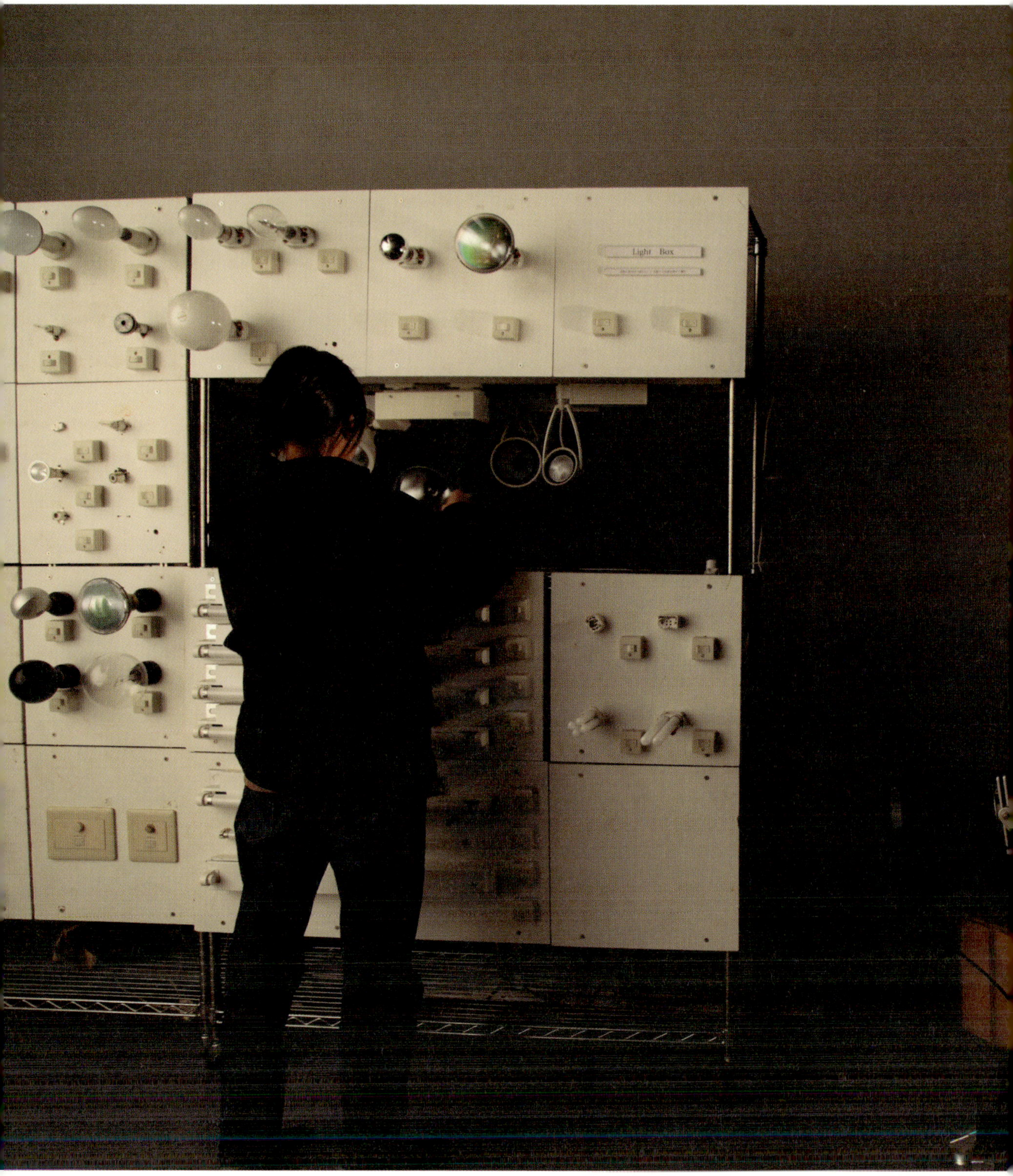
Light Box

1 武蔵美の空間演出デザイン学科には舞台演出コースがあるために、光源ボックスだけでなく、多くの種類の舞台照明用の器材を学習できる
2 最新鋭のLEDを使った照明器具もライトボックスに加えられて、岩井講師より技術講習の授業を受ける
3 様々な光源の光を点灯してその効果や性質を自ら実感し体得することが基本だ

1 In the Department of Space Design at Musashino Art University students attend class on original staging and learn about not only Light box, but other stage lighting and equipment
2 The art LEDs were also attached to Light box, then instructor Iwai gave a lesson on technology
3 Visibly studying various light sources is basic to understanding effects and characteristics of each

1	2
	3

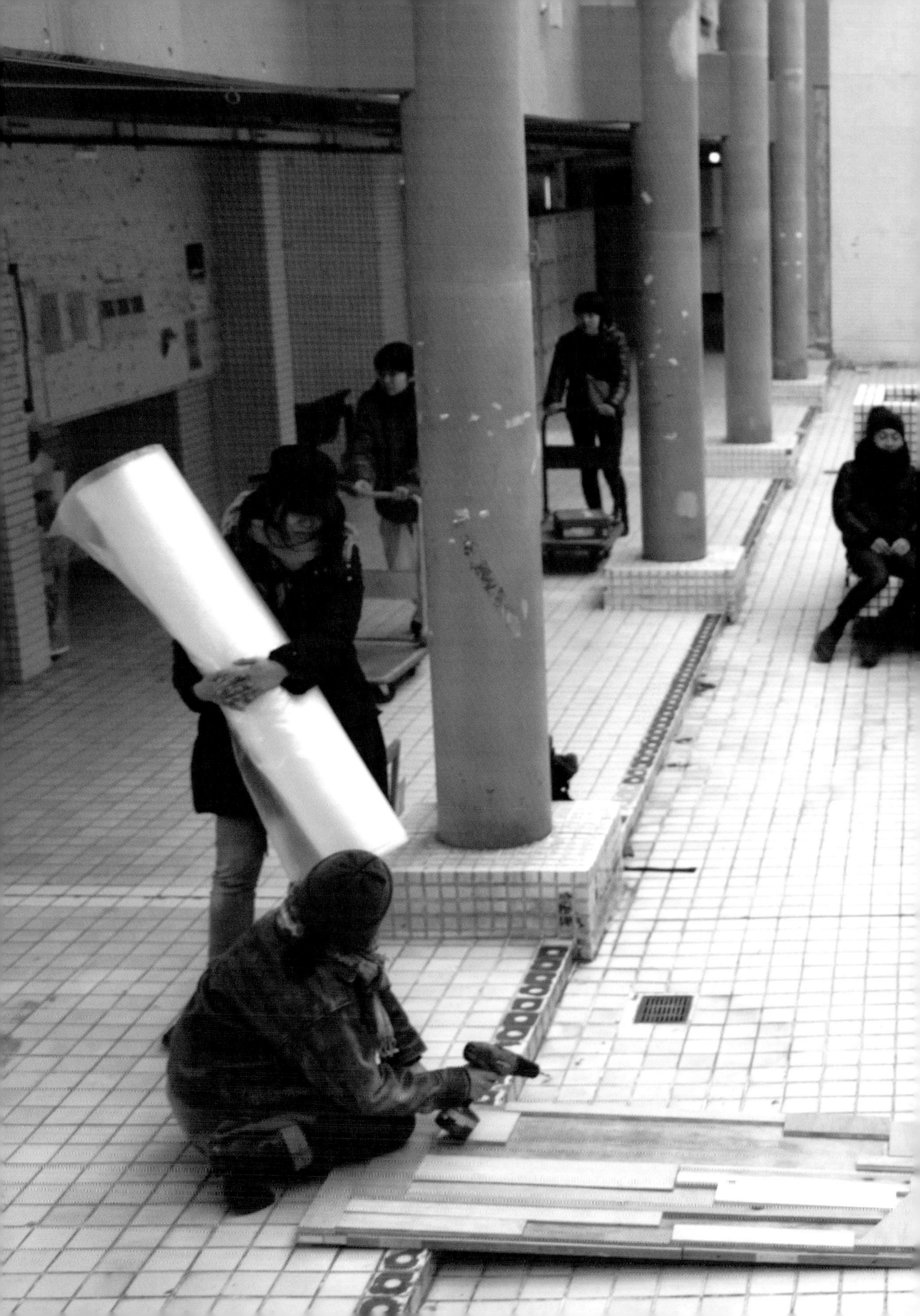

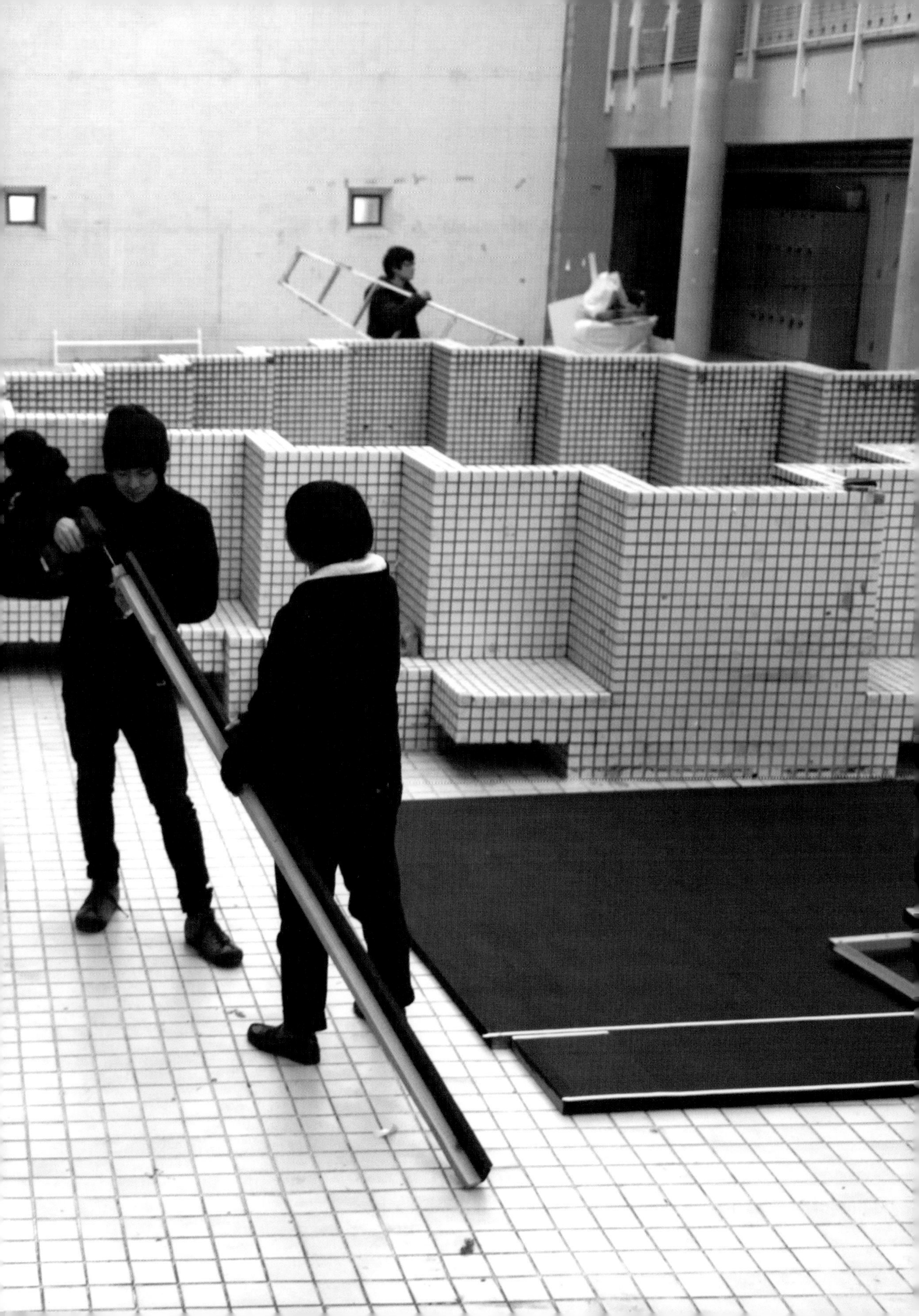

自然光に学ぶ
LEARNING FROM NATURAL LIGHT

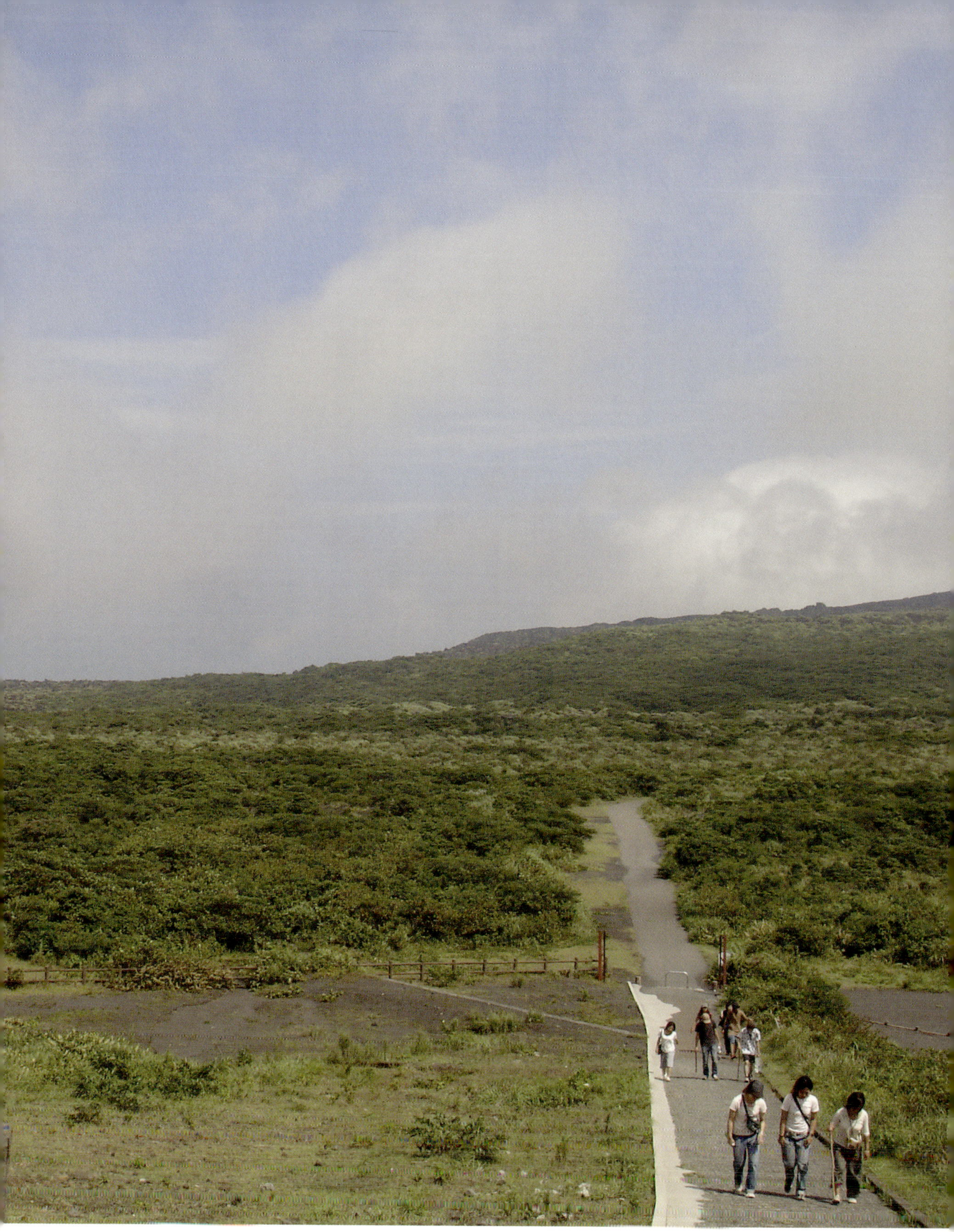

自然光に学ぶ
Learning from Natural Light

「自然光に学べ」という教訓をいつ私が発したのか明確に記憶していない。学生の頃から、近代化の影に捨て去られるものへの郷愁を抱いていたり、技術だけが先行して便利になる社会を危惧していたり、原水爆禁止運動に参加したりしていたものだから、私は漠然と自然でない生活様式に陥ることに不安を感じていたのであろう。

体内時計に変調をきたしながらも都市生活が24時間化し、安定した電力供給や照明技術の加速度的発展のお蔭で、私たちは昼のように煌々と明るい夜まで手にすることができた。しかし、夜に生活環境が暗くならないことはたいへん不自然なことなのだ。家庭の団欒を作る居間にさえ、真っ白で煌々とした真昼のような光が溢れている。オフィスのような家庭の光、これではリラックスできるはずはない。また昨今の街あかりは繁華街でもないのに、至るところでLEDによる極彩色の光がギラギラ、チカチカとうごめきあっている。これも人目を惹きたいだけの迷惑な光が多く、心休まる自然な夜景とはいえない。何が不自然な光で何が自然な光なのであろうか。このことを明確にしておかないと、照明デザインはこれ見よがしで五月蠅いだけの厄介者にもなりかねない。光のデザインを教える立場でまず、私は万人が心許す快適な光のルールについて解説する。それが「自然光に学べ」という私の第一声に繋がった。

自然光（太陽と火）の見せる技のすべてに、私たち人類は慰安されている。時々は牙をむく自然現象の脅威に身を竦ませることもあるが、それも森羅万象の摂理であれば致し方ない。光のスペクトル、光度、色温度、光の位置、陰影の深さ、移ろいの速度など、光のデザインの妙味のすべてが自然光のなせる業の中に存在する。私たちはそれこそ単純に、自然光の技を丁寧に盗むしかない。その目的のために、私たちのゼミは武蔵美の保有する教育施設である合掌造りの建物を求めて富山県の五箇山に赴き、太陽と紺碧の海の関係を知るために伊豆大島へ、そして闇夜にさんざめく星屑と月を求めて山梨県清里での合宿を繰り返した。このゼミ合宿はいつしか3年と4年の合同で、夏の終わりに毎年行われるようになった。自ら料理の腕を競い、酒を酌み交わすことで一段とゼミの輪が広がって行くことを実感した。

I cannot remember clearly when I first started telling students to learn from natural light. Since my own student days, because of nostalgia for those things modernism has left behind, misgivings about a society that gives precedence to technological advances which merely increase convenience, and participation in the movement against nuclear weapons, I have been anxious not to lapse into a lifestyle that is out of sync with nature.

Today, there is urban activity 24 hours a day, leading to irregularities in the biological clock. Thanks to a stable supply of electricity and the accelerated development of lighting technology, we have been able to make night as bright as day. However, a living environment that does not become dark at night is quite unnatural. Even the living room where the family gathers is flooded with white light. No one can relax if the household is illuminated like an office. All over town, gaudy lights using LEDs dazzle and irritate, even in districts that are not busy shopping areas. This excess of lights designed only to attract the eye do not create a restful, natural nightscape. What distinguishes natural light from unnatural light? Unless the designer knows the difference, lighting design can be simply a nuisance. As someone teaching the design of light, I first explain the rules that make light pleasant for everyone, and that led to my telling students to learn from natural light.

Humankind is comforted by the artistry of nature as evidenced in sunlight and fire. Though at times we cower in the face of a menacing natural phenomenon, we ultimately accept that that too is a part of nature. The light spectrum, luminous intensity, color temperature, position of light, depth of shadow and speed at which light fades--all the exquisite details of light design are to be found in the artistry shown by natural light. The best we can do is to carefully steal from the techniques displayed by nature. To that end, our seminar has repeatedly gone to an educational facility housed in a gassho-style structure owned by Musashino Art University in Gokayama, Toyama Prefecture, to Izu Oshima to learn the relationship between the sun and the deep blue sea, and to a lodge in Kiyosato, Yamanashi Prefecture, in search of the stars and the moon that fill the sky on a dark night. These camps with third-year and fourth-year students have become annual affairs undertaken at the end of summer where, through shared meals and drinks, the members of the seminar forge ties.

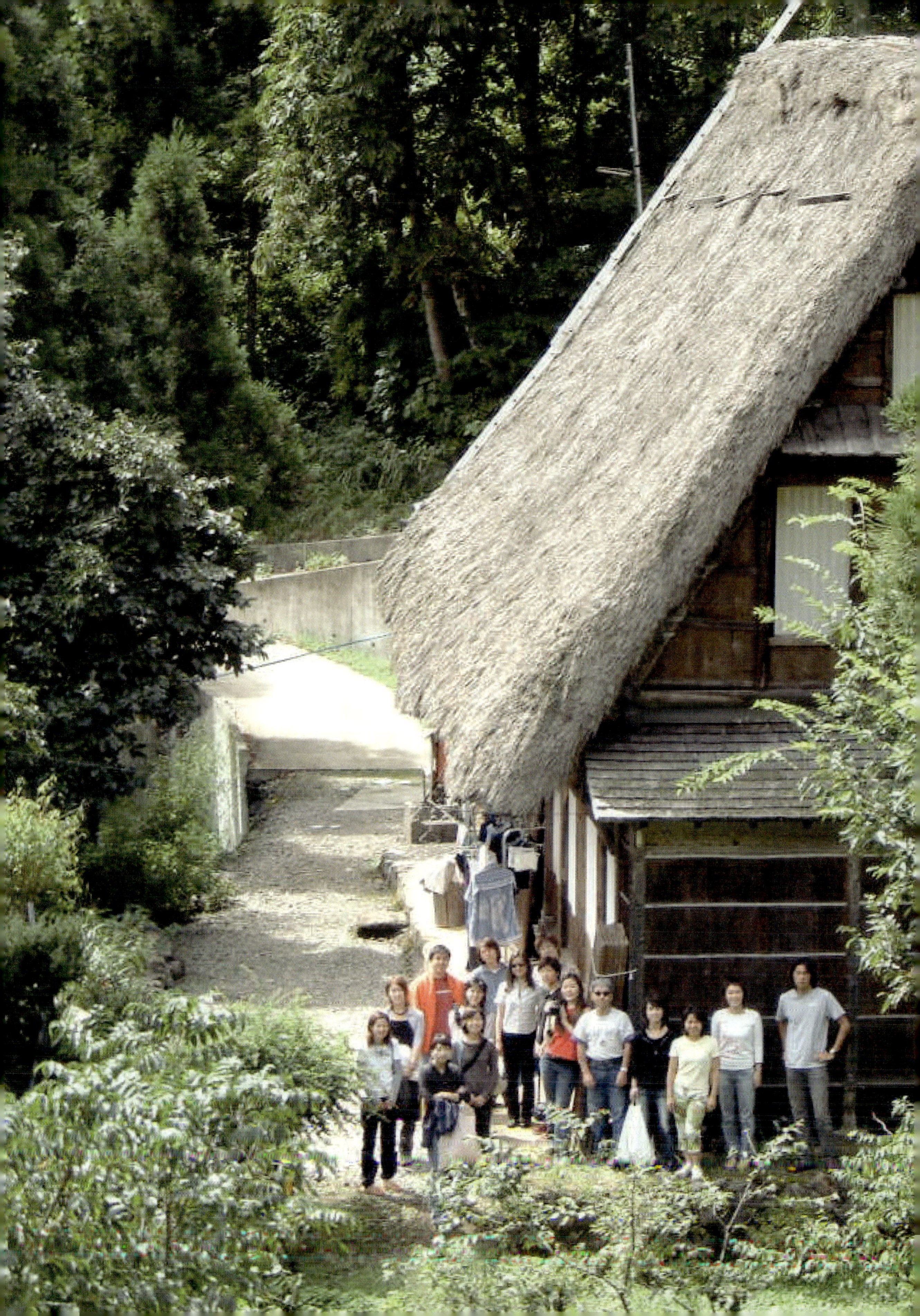

伝統民家の光 — 五箇山合宿
Light in the traditional private house Gokayama camp

武蔵美が所有する寮が富山県南砺市にある合掌づくりの民家「五箇山無名舎」である。ゼミ1期生を中心に3泊4日の自然光学習のための合宿をここで行った。都会を離れて囲炉裏を持つ保存民家に宿泊する機会を得ただけで、学生たちは興奮した。

民家の室内がどのような光環境であったのかを知るために、日中の太陽光の室内への影響を探り、夜間の囲炉裏の光や灯火による暮らしを疑似体験することにした。

ひとつ屋根の下に20人ほどで寝泊まりをするだけでも意味がある。まして面出ゼミの合宿の際には常に「心を尽くした手料理合戦」が義務付けられているので、毎晩の美食と祝宴には根性と体力を必要とする。

One of the dormitories of Musashino Art University is "Gokayama Mumeisha" located in the city of Nanto, Toyama Prefecture, which is a gasshozukuri, or an A-framed, private house with thatched roofs. Mainly first-year students participated in a camp spending four days and three nights at the dormitory to learn from natural light. Students got excited simply by having the opportunity to stay in the preserved house with a fireplace away from the city.

In order to understand the indoor light environment in the private house, students examined the effects of sunlight during the day and simulated the life with the lights of the fireplace and lamps during the night.

Life of 20 people under the same roof itself was of value. Students are constantly obliged to compete to serve one another with carefully cooked dishes in the camp of the Mende Seminar. Nightly good meals and feasts required strength in mind and body.

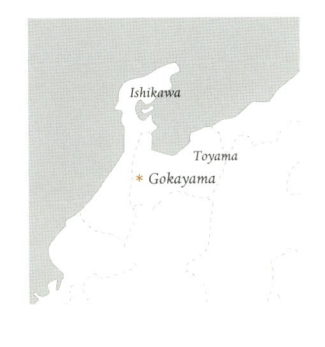

Ishikawa
Toyama
* *Gokayama*

左上｜東立面図
右上｜南立面図
下｜アクソメ図

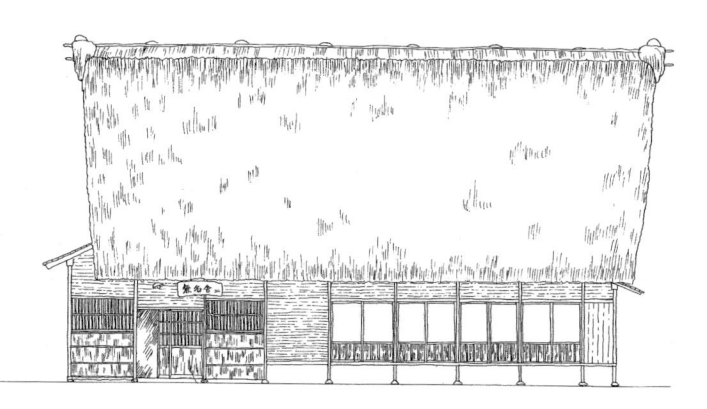

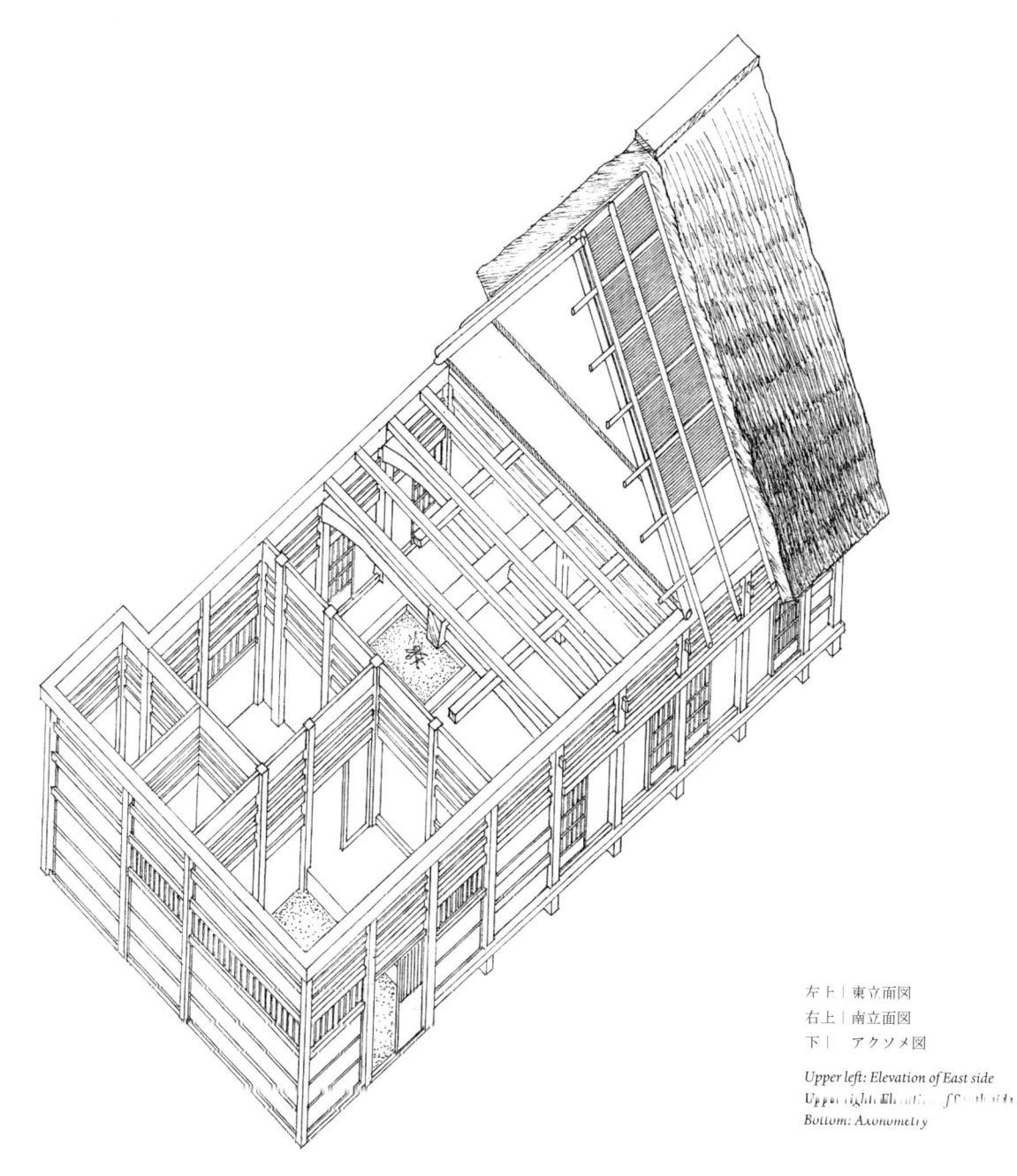

Upper left: Elevation of East side
Upper right: Elevation of South side
Bottom: Axonometry

1	2
3	4
5	6

1 面出教授が担当した朝食時のだし巻き卵。
  このレシピのために教授は専用フライパンを
  購入し自宅で練習を重ねたそうだ
2 昼間にそれぞれの担当に分かれて室内の
  光を実測した
3 外光を正しく測定することが基本
4 調査活動風景を記録班はビデオ撮影する
5 20人分の夕食は容易ではない。
  厨房で奮闘する当番学生
6 囲炉裏の火だけで会話をすると
  一体感を共有した

1 The morning Prof. Mende was in charge of breakfast
  the menu was Japanese style fried eggs
  It was rumored that Prof. Mende bought a special
  frying pan and practiced at home before the camp
2 During the day students were divided into
  teams and given a room to survey
3 Precise calculation of outside light is basic
  to surveying
4 The recording team video recorded
  the surveying in process
5 Dinner for 20 is not as easy as it sounds and
  the exertion showed in the students in charge of the kitchen
6 Conversation around an open hearth creates
  a collective sense of unity

自然光に学ぶ
LEARNING FROM NATURAL LIGHT

障子を通して自然光が柔らかく室内に入り込む。障子の配置、各建具における採光面積や光の透過率などを調べた。照度計を使った実測は初めてなので、不慣れな作業に四苦八苦した

照度の測定方法
1 水平面照度(床面照度)｜床に照度計を水平に設置して測定する。
2 床面跳ね返り照度｜床面から高さ90センチに照度計を下に向けて測定する
3 鉛直面照度｜床面から高さ90センチに」照度計を障子側に向けて測定する

Natural light filtering through paper shoji doors is soft. Based on the arrangement of shoji doors and other fittings we calculated the area of day lighting and probability of light penetration. For most of the students it was the first time to use an illuminometer and the unaccustomed work was a struggle.

Illuminance Calculating Methods
1 Horizontal Luminance (Floor lux levels): place luminance meter on floor, horizontally, and record numbers
2 Luminace reflected off the floor: hold luminance meter 90cm off the floor, facedown, and record numbers
3 Vertical luminance: hold luminance meter 90cm off the floor, facing the shoji doors, and record numbers

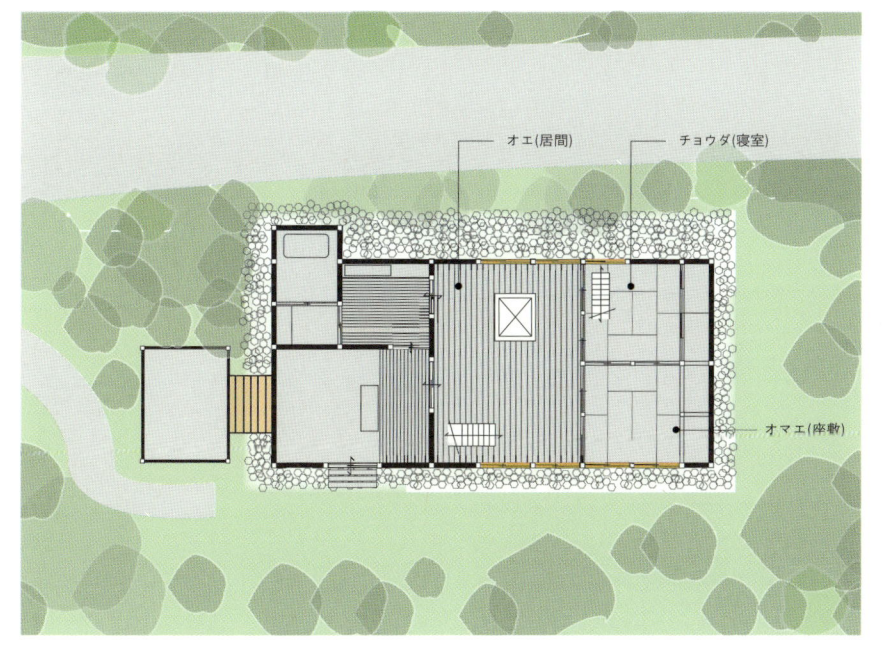

オエの床は木材で、長年かけて煤で
黒く染まり漆のような光沢を帯びている
オマエとチョウダは畳敷きである。
床材によって光の反射の様子が異なり、
部屋全体の印象にも違いを与えていた

The floor of the front room is wooden.
Over the years soot has dyed the planks black
and they have a shine like Japanese lacquer.
The two parlors are both tatami mat flooring.
Based on the floor material,
the amount of reflected light will be different
along with the entire impression of the room

南に面した障子はオエとオマエともに
同様の規格である
どちらの部屋も障子4枚で柱間が構成される

The shoji doors facing south in the front room
and first parlor are the same specifications.
Both rooms are composed of four doors each,
in between the column spacing

素材: ガラス (凸凹あり)
採光面積: 3006cm² 23.2% (1枚あたり)

素材: 和紙
透過率: 29.5%
採光面積: 5505cm² 40.1% (1枚あたり)

自然光に学ぶ
LEARNING FROM NATURAL LIGHT

測定方法
オエ（居間）と、オマエ（座敷）とチョウダ（座敷）を合わせた部屋の2カ所で、照度測定をした。障子から室内に向かって900ミリごとに1.床面照度、2.床面跳ね返り照度、3.鉛直面照度を計測する。障子を開けた時と閉めた時の両方を記録した。

測定結果
障子を開いた時には約3メートルの距離で急に照度が低くなり、部屋の奥に闇のたまりができたように見える。オエでは光が床材に反射するために奥まで光が取り込まれていた。
障子を閉じた時には障子際に光が溜まっているように見える。1メートル以内の範囲で250ルクスだが奥に行くにつれて100ルクス以下に下がって行った。

水平面照度（床面照度）
Horizontal luminance

床面跳ね返り照度
Luminance reflected off the floor

鉛直面照度
Vertical luminance

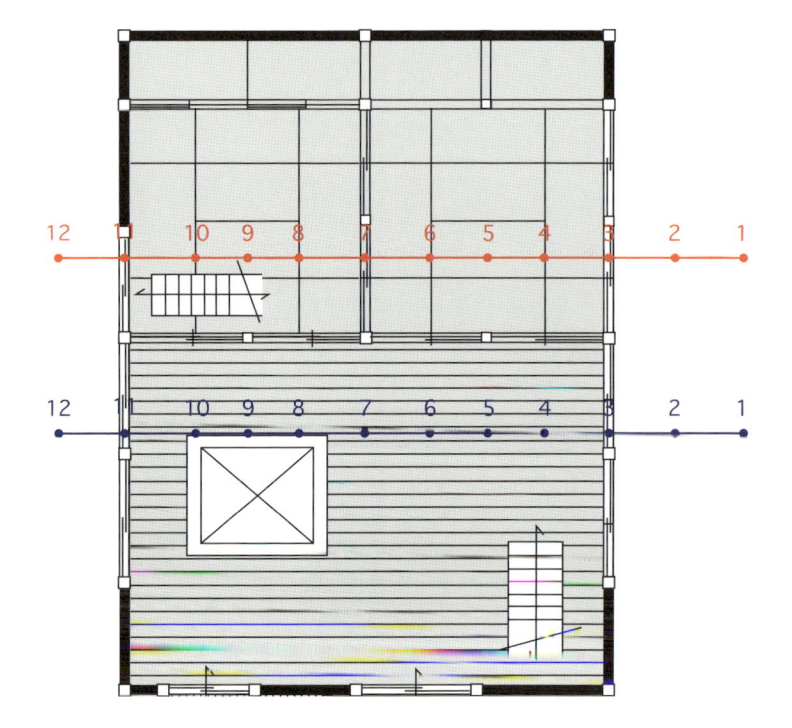

左｜照度の測定位置
　　90cm間隔で上記3つの照度を測定
右｜照度計は数値で入射する光の量を捉えるものだが、見ための印象も大切なので、スケッチを残すことにした。3人がかりで、A1サイズのケント紙に数時間かけて描いた。スケッチで記録することは時間もかかるが、体験とした光の印象を伝えるには実測照度以上に効力を発揮する

L: *Calculating spots of luminance calculate*
　Three ways of luminance above, at intervals of 90cm
R: *The luminance meter captures the amount of light coming into the room in numbers, but how we see the light is also very important. Therefore, we broke into groups of three to sketch the rooms on A1 size drawing paper, taking over several hours. Sketching takes time, but to communicate the impression of the light, this method might be much more effective than recording numbers*

Calculating Methods:

In the Japanese style front room and two adjoining parlors, we calculated lux levels in two areas. From the shoji doors 900mm inside we measured 1 floor luminance, 2 luminance that reflects off the floor, and 3 vertical luminance. We recorded the data with the doors both open and shut.

Calculating Results:

When the shoji doors are open, about three meters into the room the lux levels immediately drop and the back of the room seems to be very dark. In the front room light reflects off the floor material, so light reaches the back of the room. When the shoji doors are closed, light seems to linger around the doors. Within one meter from the door we recorded 250 lux, but moving toward the back of the room lux levels dipped below 100 lux.

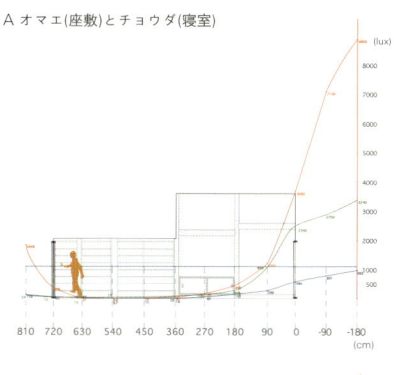

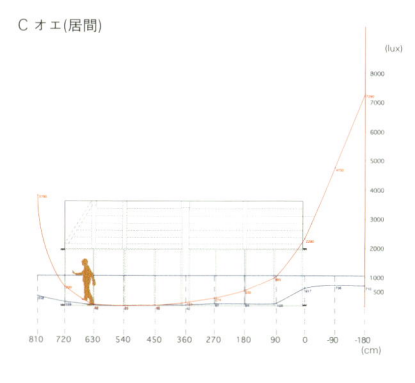

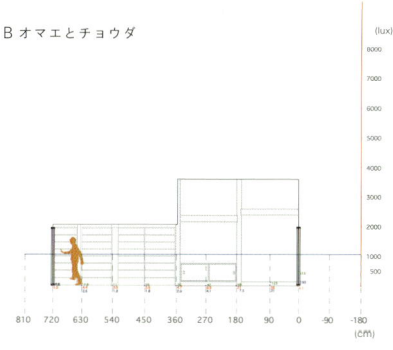

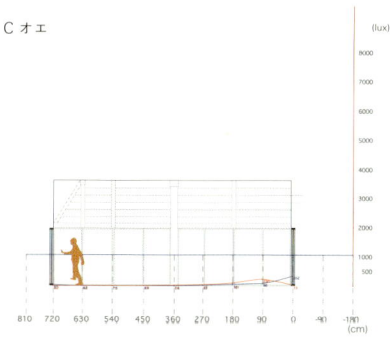

緑と海の合宿
清里合宿・大島合宿
Annual camps with green and sea in Kiyosato and Ohshima

「自然光に学ぶ」というゼミだから、自然光が純粋に味わえる場所に行って寝泊まりし、酒を酌み交わして気心を通わせるべきだと考えた。

　ゼミは3年生の後期9月から始まり4年の通期までの約18か月で終了する。ゼミ室は3年と4年が共通して使用するので彼らは一心同体の関係を強いられる。新3年ゼミ生を迎える際の儀式としても合同ゼミ合宿は機能した。山梨県清里には山の緑が、東京都大島には海の青がある。それを交互に訪ねるように計画した。

　清里にも大島にも美しい夜が訪れ、闇を背景にしたオレンジ色の灯火や、恵まれた時には危うい月あかりにも酔いしれた。学生たちは束の間に、自然光と友情を深めることになる。

I thought it necessary for us to stay and drink together for communication in places where we can genuinely enjoy natural light since we are in a seminar for learning from natural light.

Seminars start in September in the second semester of third year and end in about 18 months at the end of the full fourth year. The seminar room is shared by third- and fourth-year students, so they are forced to act as one. The joint camps also served as a ceremony for welcoming new third-year students. Kiyosato, Yamanashi Prefecture is endowed with green mountains and Ohshima, Tokyo with blue seas. The camps were planned to provide both to students alternately.

Beautiful nights came to Kiyosato and to Ohshima. We were enchanted by orange lights against darkness and, when we were lucky, by unstable moonlight. Students enjoyed natural light for a fleeting moment.

各年の合宿地 Places of annual camps	清里 Kiyosato	大島 Oshima	その他 Others
2002			* 五箇山 Gokayama
2003	*		
2004		*	
2005	*		
2006	*		
2007		*	
2008	*		
2009		*	
2010	*		
2011			* 箱根 Hakone
2012	*		

* Kiyosato, Yamanashi

* Oshima, Tokyo

自然光に学ぶ
LEARNING FROM NATURAL LIGHT

緑の合宿 ── 清里
Green Annual Camp ── Kiyosato

武蔵美の清里寮は屋外でBBQを楽しむ施設や、夜通し騒いで飲み明かせる離れがあり、虫の音や月あかりに囲まれ、時間を忘れてデザインの話や時に恋愛の話までする。立派なキッチンを使って特別料理を作り、グループ毎のおもてなし合戦を行う。昼間は牧場で体を使い夜にはとことん飲む。これが清里スタイルとなって行った。

At the Musashino Art U. Kiyosato facility there is an annex building for BBQs and late night parties, and it also happens to be surrounded by the sounds of nature and moonlight. This where we are able to forget about time and talk endlessly about design and sometimes of love and life. The annex is also equipped with a full kitchen, so we have a cook-off. Each group has to prepare and serve a special dish. After a full day of running around a local ranch, we relax with great food and drink. This has come to be Kiyosato style.

1　緑のなか目一杯、身体を動かす
2　毎晩、担当者たちより工夫を凝らした夕食が供される。これもまた面出ゼミならではのグループワークである
3　キャンドルの朧げなあかりのもと、ゼミ生の親睦は深まってゆく
4　木陰のなか、ついウトウト眠り込んでしまう面出教授
5　花火大会。合宿当初はぎこちなくも、このときゼミ生達はすっかり打ち解けている
6　深く、さわやかな木漏れ陽を浴びながらの贅沢な朝食

1　Fully moving your body in greenery
2　Participants take turns in serving elaborately prepared supper every night.
　　This is a part of the group work typical of the Mende Seminar
3　Seminar members gradually develop friendly ties with one another under dim candlelight
4　Professor Mende drowsing in the shade of trees
5　Fireworks event. Seimar members, awkward at the beginning of the camp, are now totally open to one another
6　Luxurious breakfast under the brisk sunlight falling through thick trees

自然光に学ぶ
LEARNING FROM NATURAL LIGHT

0.2ルクスの月夜に佇む
Standing in 0.2-lux moonlight

2012年9月1日、清里での面出ゼミ最終合宿の夜、私たちは満月に佇んだ。

灼熱の太陽が100,000ルクス、オフィス照明が1000ルクス、住宅照明が100ルクス、そしてこの清里の月光は0.2ルクス。

「どうして0.2ルクスがこんなに明るいの…」

「この微かな雰囲気に吸い込まれそう…」

「これは太陽からの反射光なんだ…」

私たちにとって目からウロコが落ちる瞬間だった。

月の光は私たちに人間と光の原点を教えてくれているかのようだった。

We stood under a full moon in Kiyosato on the night of September 1, 2012, the last day of a Mende Seminar camp. The luminance of the moonlight in Kiyosata was 0.2 lux while the luminance of a blazing sun, office lighting and house lighting was 100,000, 1,000 and 100 lux, respectively.

"Why is a 0.2-lux light so bright?"

"I am likely to be absorbed in the subtle air."

"This is a reflection of sunlight."

At the moment, we suddenly understood the truth. The moonlight seemed to teach us the origin of people and light.

99

海の合宿──大島
Sea Annual Camp ── Ohshima

海の合宿は潮騒の聞こえる海辺のキャンプ地に陣をはる。海辺に降りればすぐに透明度の高い海が待っている。夕食に貢献しようとして素潜りをする者や磯釣りに励む者もいるが、褒められたことはない。全員で火をおこしたり飯を炊いたり。台風や天候だけが気がかりだが、たっぷり自然の中で様々な光や陰影に触れることができる。

At the sea training camp we pitch a tent at a seaside camp ground within earshot of the sounds of the surf. Once down on the beach, only the clear blue waters await. To contribute to dinner some students skin dive or fish in the surf, but their effort is never rewarded. All at the camp work to start a fire and make rice and dinner. Every year we are anxious over the weather or incoming typhoons, but in this natural surrounding it is an opportunity to be exposed to various light and shadow.

1　海を見ながらの朝食。雄大な波の音を
　　感じながら大島の朝は始まる
2　自ら自然の一部となりながら海で遊ぶ
3　三原山の登山。時に晴れ、時に霧に包まれ、
　　刻一刻と変化する光環境を体験する
4　ここでも料理当番の出番。海の幸をいかした
　　料理がつくられる
5　大島の夕暮れ。すべてがオレンジに染まる瞬間
6　月明かりのもと、我を忘れるかのように
　　楽しむ花火大会

1　*Breakfast while enjoying the view of the sea. Morning starts in Ohshima as you feel the sound of splendid waves*
2　*Enjoying oneself while becoming a part of nature*
3　*Climbing Mr. Mihara. The weather was fine at one time and foggy at another. We experienced ever changing light environment*
4　*Taking turns in serving meals also in this camp. Dishes are made making use of the fruits of the sea*
5　*Dust in Ohshima. A moment when everything turns orange*
6　*We enjoyed fireworks as if we lost ourselves under the moonlight*

自然光に学ぶ
LEARNING FROM NATURAL LIGHT

101

オーロラ観測隊
Aurora observation team

2005年2月28日〜3月6日まで、ゼミ生有志を募ってアラスカ・フェアバンクスにオーロラ観測のための調査旅行に出た。何しろマイナス30°程度の外気に長時間さらされながら三脚にカメラを据えて劇的なオーロラの出現を待つのだから、出発前の準備が大変だった。

　私たちの身体はもとより、カメラやバッテリーが機能するための防寒対策に時間をかけた。観測は4晩続いたが、劇的なオーロラの変化（ブレイク）に出会ったのは一度だけ。南北両方の低い空から同時に竜のようなオーロラが立ち上り、私たちの頭上で交差した。その間わずかに30秒程度だったが、私たちは十分興奮した。

We made an aurora observation tour to Fairbanks, Alaska with some seminar member students on February 28 through March 6, 2005. We were supposed to wait for the appearance of aurora while being exposed to outer air at minus 30 degrees with cameras installed on tripods. Laborious preparations were made before departure. Time was spent on cold protection not only of ourselves but also of cameras and batteries to ensure proper function. We encountered a dramatic outbreak of aurora only once during four-night continual observation. Aurora emerged like cascades from low southern and northern skies and crisscrossed each other overhead. The show of only about 30 seconds was enough to excite us all.

低い位置のオーロラを背景に逞しく佇む6名の観測隊員
The six member survey team stands diligently before a low-forming aurora

1 飛行機の中でも食欲旺盛な人と睡魔に襲われる人と…
2 ホテルを出て朝日を浴びるチナ川あたりを就寝前に散歩する
3 観測に向かう前に広大なスーパーマーケットで食料の調達
4 深夜の3時頃にホテルに戻り、それからがまた楽しいひと仕事。撮影した映像をPCで整理してデータを付け加える

1 *After returning to the hotel at 3 AM the exciting second half of the night awaits. Team members download images to the computer and input data*
2 *Before our night of surveying begins, shopping at the local supermarket*
3 *Before heading to bed, a walk along the Chena River in the morning sun*
4 *Back in the air, those with a hearty appetite and those soundly asleep*

installation系

みとす
光をまとう　ウェアラブル・ファッション
身につける光
撮影?
Open City

ICC

音と色（光）
Light & Sound
ヤス
無響堂

Product系
Furniture

2011
201?

ジン
貝、魚のウロコ
照明器具
試作品 4コ
9/1?

Material系

なおり？
言葉+映像+空間
詩
映像の中に入り込む
手を動かしたい

ロウを使って…
キャンドルじゃないものを作る
(ハスの花)

年輪・流木など
木っぽい素材
養蚕と

タクト
これからどうなるか
人体と光…の関係
人が空間に入ることで…
ものが光る。

Research系
研究所

パイ
光をたくさん収集する
光の標本
光のボキャブラリー/S.D

photo
D

Architecture系．
interore．

philosopy系

もうさき．まい
まけwwwの空間
ろい，グラデション
の素室
交句
部え

ひかる
光のエッジ
レンズ効果
プロジェクション
車のヘッドライト

生命について
生きるって何が
おきまっ
言葉

Environment系．

ng系
あやね
エッチング
銅版
写真と思う生

レンズを通して見た質感．

ジオラマ
粘土でつかう
しめり

宮橋
芳醇清酒
登録商標
東北に酒あり
エイセン
沖正宗
芳醇無比
浜田株式会社
東光
清酒
七賢
芳醇清酒
松美酉
清酒
金井謹醸
伝心
天狗舞
芳醇清酒
文政年間
車多酒造
鹿野酒造
日本城
長久
名聲秀天下
尾崎酒造

闇を知る
KNOWING DARKNESS

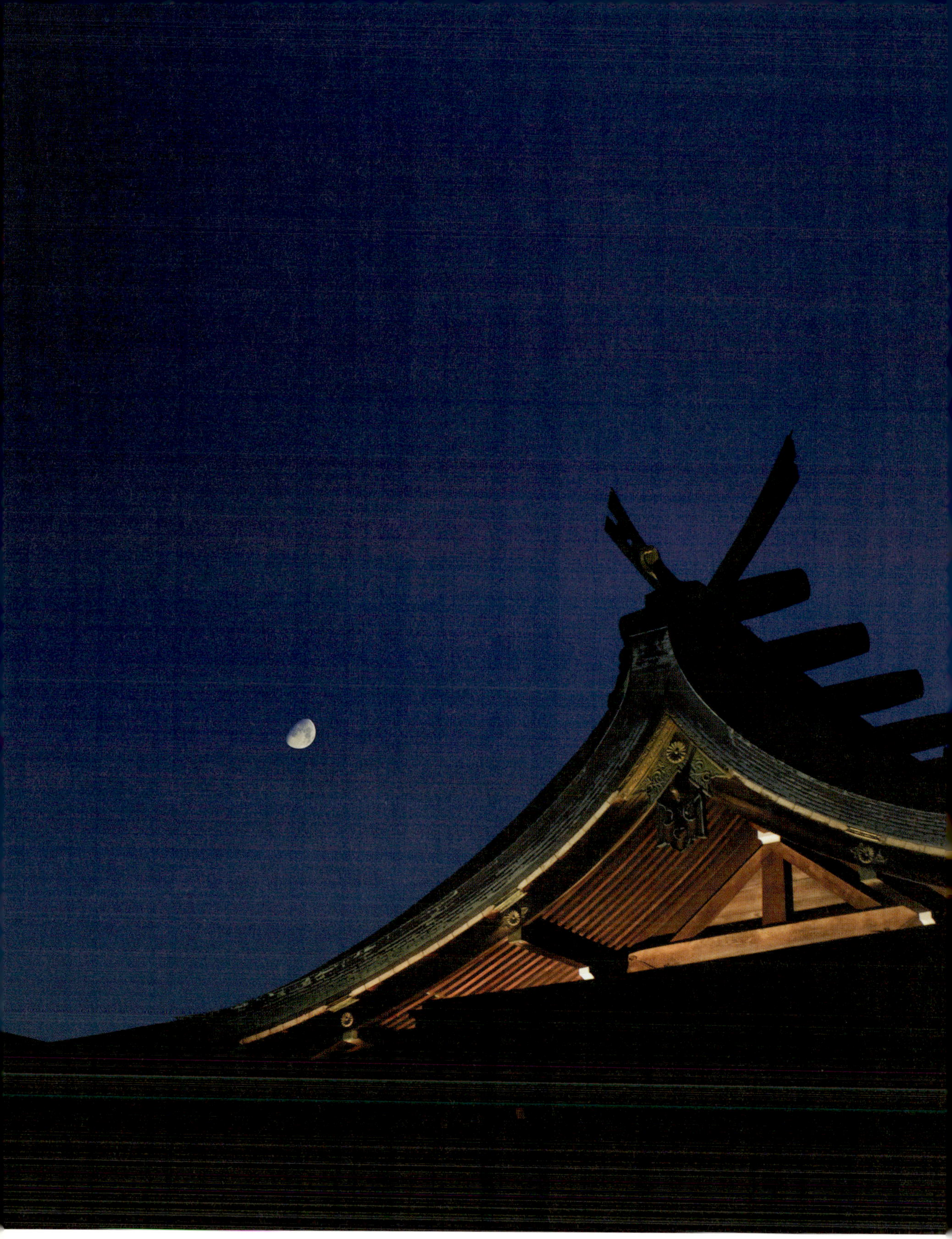

闇を知る
Knowing Darkness

照明デザインの原点は闇であり陰影だと考えている。それは禅問答に似て光が先でも闇が先でも良いのだが、とりわけ戦後の60年で私たちは闇を葬り陰影を排斥してきたので、その近代化へのアンチテーゼとして、私は闇や陰影が照明デザインの原点なのだ、という立場を学生に主張してきた。

闇に一条の光が差し込み、1本の蝋燭が灯されることにより初めて、光のデザインは視覚化され客体化する。つまり、太陽光に支配される日中の屋外には光のデザインなど成立しない。灼熱の太陽にさらされるイスラムの都市では、建築様式を工夫して太陽の直射から逃れて安堵する闇や陰影をデザインすることに腐心する。光ありきではなく、闇から出発するしかないのだ。

しかし、私が海外で講演する際に、美しい闇（Beautiful Darkness）という言葉を使うことがあるが、その時に必ずといっていいほど「美しい闇とは何か」という質問が返ってくる。私にとって闇とは僅かな光が登場するための一瞬のセットであるのに、欧米人にとって闇は完璧に受け入れがたい困ったものであるらしい。そのような質問が来る度に私は、ひたすら日本人が僅かな明るさを愛で、柔らかな陰影を慈しみ、畏怖を覚えながらも闇に精神的な拠りどころを感じていることを説明する。そして洗練された光のデザインを希求する時に、闇なしで光が成立しないことを諭すのみである。「闇は美しいか」と尋ねられて、「闇は光なのです」と答えたいところだが、私はそこまで哲学者を気取るつもりはない。

私は学生たちと一緒に、時々「完璧に暗転できる部屋」を作り出す。窓からの自然光を何重にも丁寧に遮光し、両目を見開いていても何の視覚情報もなくなるような環境を創り出す。そこで遭遇する浮遊した時間の感覚、それが大切なのだ。視覚情報以外の音や匂いや風の向きなどに敏感になることが分かる。東京には闇を感じる屋外空間はもちろん皆無だ。しかし限りなく闇に近い都心の屋外環境を発見した。原宿駅に隣接する明治神宮の杜だ。私たちのゼミはこの唯一残された都会の闇を相手に、「闇のライトアップ」と命名した妙な光の実験を試みた。明治神宮の夜は何かが潜むがごとく芸術的に暗かった。

I believe the starting point of lighting design is darkness and shadow. That may seem like a dialogue, and in truth the starting point could just as well be light. However, as we have banished darkness and shadows particularly in the 60 years since the end of World War II, I have made it a point to tell students that the starting point of lighting design is darkness and shadow and in that way suggest an antithesis to modernization.

With the introduction of a ray of light or a lit candle in the darkness, the design of light is made perceptible and externalized for the first time. That is to say, there can be no design of light outdoors in the daytime when everything is illuminated by sunlight. In Islamic cities exposed to the scorching sun, great effort is put into devising architectural styles so as to design darkness or shadows that provide relief from direct sunlight. In such cases, one can only start from darkness.

However, whenever I describe darkness as "beautiful" in a lecture overseas, I am inevitably asked, "What do you mean by beautiful darkness?" Though darkness for me is a condition preliminary to the appearance of light, to Westerners, darkness seems to be completely unacceptable and problematic. Every time I am asked such a question, I explain that we Japanese appreciate faint lights, love soft shadows and are psychologically comforted by darkness even as we are awed by it. I can only try to persuade the questioner that darkness is essential to light and to the refinement of the design of light. Asked, "Is darkness beautiful?", my first instinct is to answer, "Darkness is light," but that seems too much like something a philosopher with a taste for paradox might say.

Occasionally, I and my students construct a room that can be completely blacked out. Natural light from windows is carefully screened by layers of some material, and an environment where one receives no visual information even with eyes wide open is created. In the room, one experiences a sense of time floating. Sensitivity to non-visual information such as sounds, smells and the direction of air movement is heightened. There is of course no outdoor space in Tokyo where one is completely in the dark, but I have discovered a place in the middle of the city that approaches that condition—the woods of Meiji Shrine next to Harajuku Station. There, our seminar tried an interesting experiment with light which we called "Light up the Darkness." Meiji Shrine at night was so artistically dark that something seemed to be hiding there.

明治神宮の闇に学ぶ
Learning from darkness in the Meiji Shrine

2004年末に表参道アカリウムという照明演出イベントが表参道ケヤキ道で行われたきっかけで、明治神宮の中島精太郎宮司にお目にかかった。「ここ明治神宮には芸術に通じる美しい闇が残されている」と語られた宮司の熱意に強く感動し、私たちは期間限定の夜間参拝のための照明計画を快くお引き受けした。

　私は「明治神宮の夜の絵はがきづくり」と「闇のライトアップ」という2つのプロジェクトをゼミ生総動員で立ち上げた。こんな素晴らしい闇と対峙する経験は二度とない。しかし「闇の芸術性」をどのように表現しようか？原宿に眠る闇を相手に、現場主義のゼミ生たちは目を輝かせてこれに取り組んだ。

When Omotesando Akarium, a lighting dramatization event, was held along the Omotesando Avenue lined with zelkova trees, I had an opportunity to meet Mr. Seitaro Nakajima, head priest of the shrine. We were deeply moved by priest's words, "Beautiful darkness remains in the Meiji Shrine that may be regarded as art." We willingly accepted his request to plan lighting for nighttime worshippers for a limited period of time.

　I launched the "Meiji Shrine night picture postcards" and "darkness light-up" projects with the participation of all seminar members. There would be no other chance to face up to such wonderful darkness. How should the artistic character of darkness be expressed? Realistic seminar students tackled the assignment with sparkling eyes against the darkness lying in Harajuku.

闇を知る
KNOWING DARKNESS

夜の絵はがきづくり
Making "Night picture postcards"

明治神宮の売店にはもちろん絵葉書が販売されている。しかしその絵葉書は全て昼間に撮影された写真によるもので、夜景は1枚もなかった。「明治神宮の美しく神聖な夜景を絵葉書にしよう」という企画が決定した。通常では午後5時には閉門し参拝を終了しているが、将来的には残照の残る夕暮れ時に参拝していただきたい、と考えた。

　木造の和風建築は石造りの欧風建築とは異なり、建築内部から漏れてくる暖かいあかりの風情が大切である。外からの投光照明を極力なくして、光の量ではなく質にこだわったライトアップを目指して奮闘した。結果的に10枚ほどの夜の絵葉書を作成して明治神宮に奉納し、現在そのうちの数枚が使用され販売されている。

Postcards were naturally sold at the shops in the Meiji Shrine. All were, however, used pictures taken during the daytime. No night views were included. A decision was made to "develop postcards of divine night views of the Meiji Shrine". Gates were generally closed to worshippers at five in the afternoon but it was hoped that worshippers would visit at dust in the future when afterglow was lingering.

　For Japanese style architecture made of wood, unlike for stone-built European counterpart, the taste of warm light pouring from inside is precious. We made strenuous efforts to seek light-up focused not on the quantity but on the quality of light while minimizing the lighting projected from outside. In the end, we developed and dedicated about ten postcards with night views to the shrine. A few of them are on sale at present.

1 本殿で仕事をする女性は白い装束を
  つけることが義務付けられた
2 学生が一人ずつ照明器具の役割を担った
3 必要な輝度を測定する
4 屋根に登り南神門の中に照明を仕込む
5 各種の照明装置で点灯実験が行われた
6 深い森の中にも神秘的な光を仕込もうとしている
7 昼間には各担当の配置が確認される
8 プロのカメラマンの指示に従って微妙な
  光を調整する
9 参道に青いフィルターをかけた光が与えられた

1 Ladies working at the main shrine are required to
  wear white costumes
2 Each student was assigned the duties
  of a light fixture
3 Calculating the necessary amount of luminance
4 Climbing up on the roof of the south gate
  to possession fixtures
5 Lighting experiments using various types
  of light fixtures
6 Adding mystical light to the dark forest
7 Checking the position of fixtures during the day
8 A professional photographer directed students
  in coordinating sensitive lighting details
9 Fixtures with blue filters light the main approach

シンメトリカルな鳥居の造形に対して、背後の樹木の枝ぶりなどを繊細に照明しながら、心に残る夜の絵にしようとした。鳥居に対する照明も、ハイライトと影の部分のコントラストを微妙に変化させている

We tried to create an unforgettable picture by finely lighting the foliage behind Torii of symmetrical design. Lighting for Torii was slightly varied to change the contrast in highlighted and shaded areas

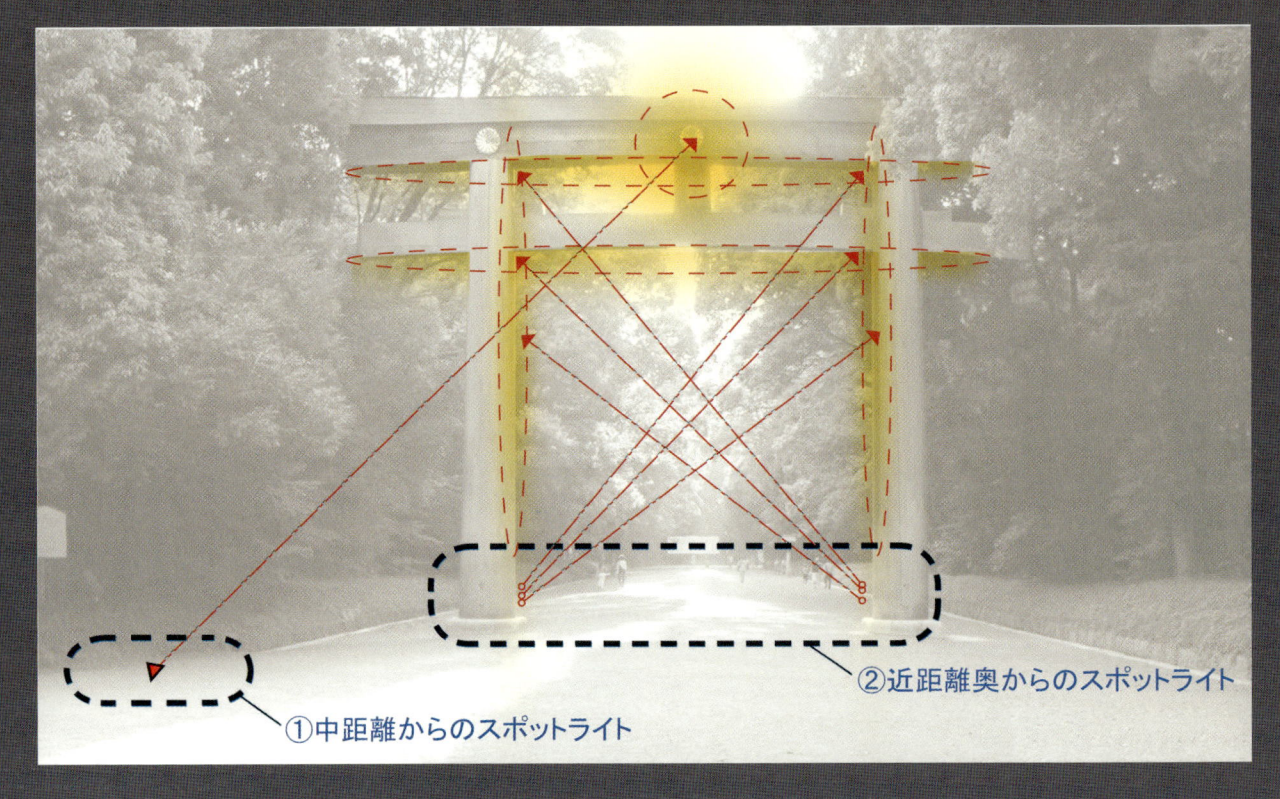

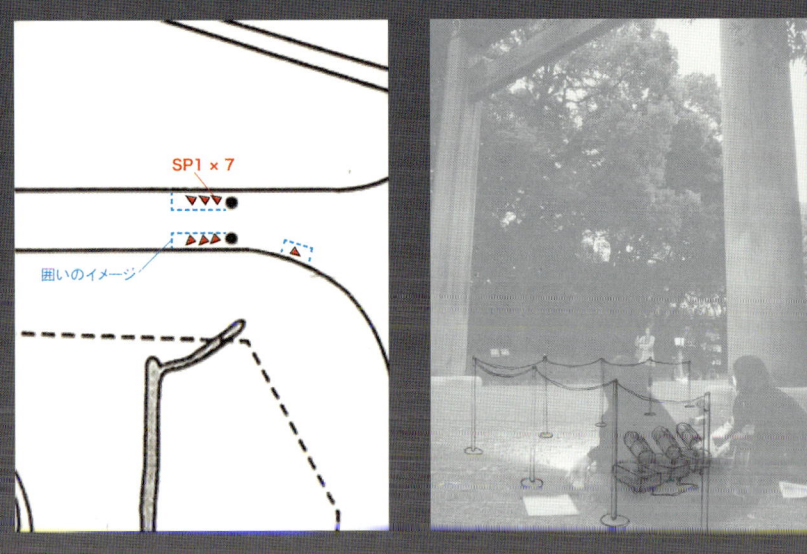

①中距離からのスポットライト

②近距離奥からのスポットライト

SP1 × 7

囲いのイメージ

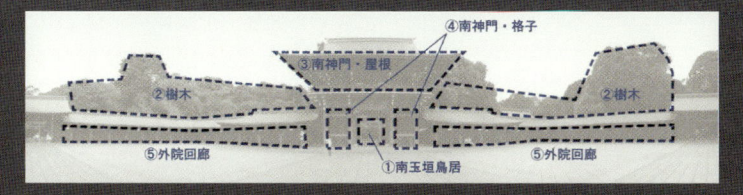

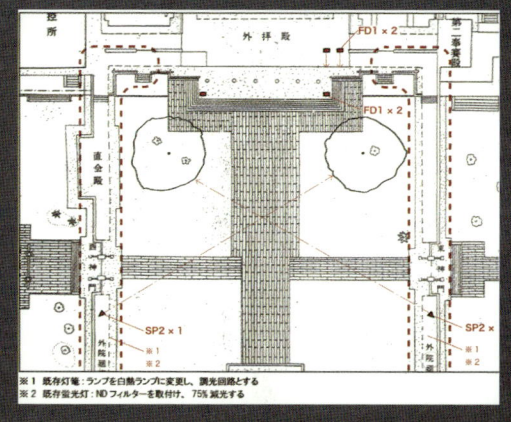

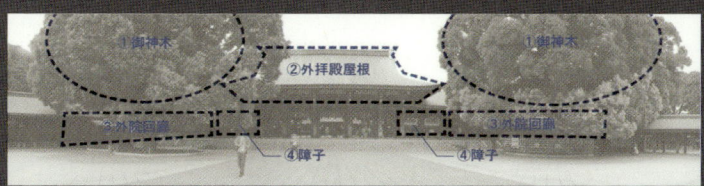

和風の佇まいに大切なのはエレベーションの光だ。煌々と照らさずに自然な感じを出すための立面上の検討に時間をかける。平面図には最終的な照明器具の配置（配当図）が描かれる

The most import element when dealing with Japanese style architecture is the elevation lighting. The elevation shouldn't be brightly lit, but should seem more natural and this takes time to study the detail of the entire elevation. Finally, the layout of the lighting fixtures is drawn on the plans. (Fixture Layout Diagram)

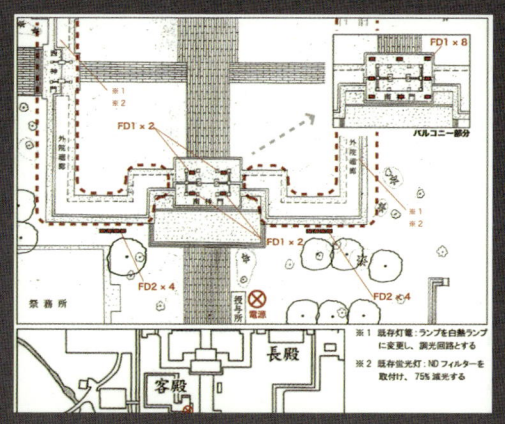

闇を知る
KNOWING DARKNESS

奉納された酒樽のディスプレイ
背景の緑を酒樽の表情と同時に演出した

Display of sake barrels dedicated to the shrine :
The greenery in the background was also dramatized
as well as the expression of sake barrels

橋から見える深い森
森の奥深く神秘的に霞んだ情景を創ろうとした

Thick forest seen from a bridge :
We tried to create mysterious and misty scenes deep in the forest

大鳥居
鳥居をくぐりたくなるような奥の景色も重要だった

Big torii :
The landscape in the far was also important that tempted people
to get under the torii

南神門
建物の内部から溢れ出るあかりを大切に扱った

Nanjin-mon gate : We carefully handled the light overflowing the building

外拝殿
手前に構える2本の大樹もフォーカルポイントだ

*Gaihai-den, or external hall of worship :
Two tall trees in the near were also focal points*

先ずはイメージをスケッチにしてみる。1枚の絵葉書になるためには、切り取ろうとする景色が10枚の完成すべき絵葉書の中でどのような個性を発揮すべきなのかを考える。

夜の絵葉書を創るという行為は、多様な性格を持つ光源や照明器具を絵具や絵筆に例えて、心にしみる「光の絵」を描くようなものである。

昼間の景色に対して、夜にどれだけ印象的な情景を描けるかという挑戦だ。

We first developed a sketch of an image. In order to create a postcard, we considered how it should exhibit its individuality among a ten landscape postcards. Creating a night picture postcard is something like drawing a touching "picture of light" with paints and paint brushes instead of light sources and lighting equipment of diverse characteristics.

The challenge was to draw impressive scenes at night as compared with daytime landscape.

闇のライトアップ
Light-up the darkness

夜の絵葉書プロジェクトが終わった次の年2008年に、明治神宮御社殿復興50年記念「アカリウム」というプロジェクトが行われた。1週間ほどの限定期間、多くの方々に夜間参拝をしていただき、神宮に宿る闇の素晴らしさを満喫していただこうという企画だ。環境に優しいエコロジカルなあかりの演出をテーマにした。

JR原宿駅に隣接する原宿門から入場し南参道を抜けて南神門、そして本殿へと続く約15分ほどの道のりを、場の連続性を考慮した照明デザインを施した。このプロジェクトはLPAが全体の計画を担当し、面出ゼミは主に北参道と西広場の計画を責任担当した。

In the year following the development of night picture postcards, the "Akarium" project was implemented commemorating the 50th anniversary of the reconstruction of the pavilions of the Meiji Shrine. The goal was to have numerous nighttime worshippers enjoy the magnificent darkness that resided in the Shrine during a limited period of about one week. The theme was to dramatize environmentally friendly, ecological light.

Lighting was designed for a 15-minute walking distance from the Harajuku gate adjacent to Japan Railways Harajuku Station to the main shrine via south approach and Nanjin-mon to maintain continuity. The Lighting Planners Association (LPA) was responsible for overall project planning. The Mende Seminar was held responsible mainly for the planning of the north approach and west square.

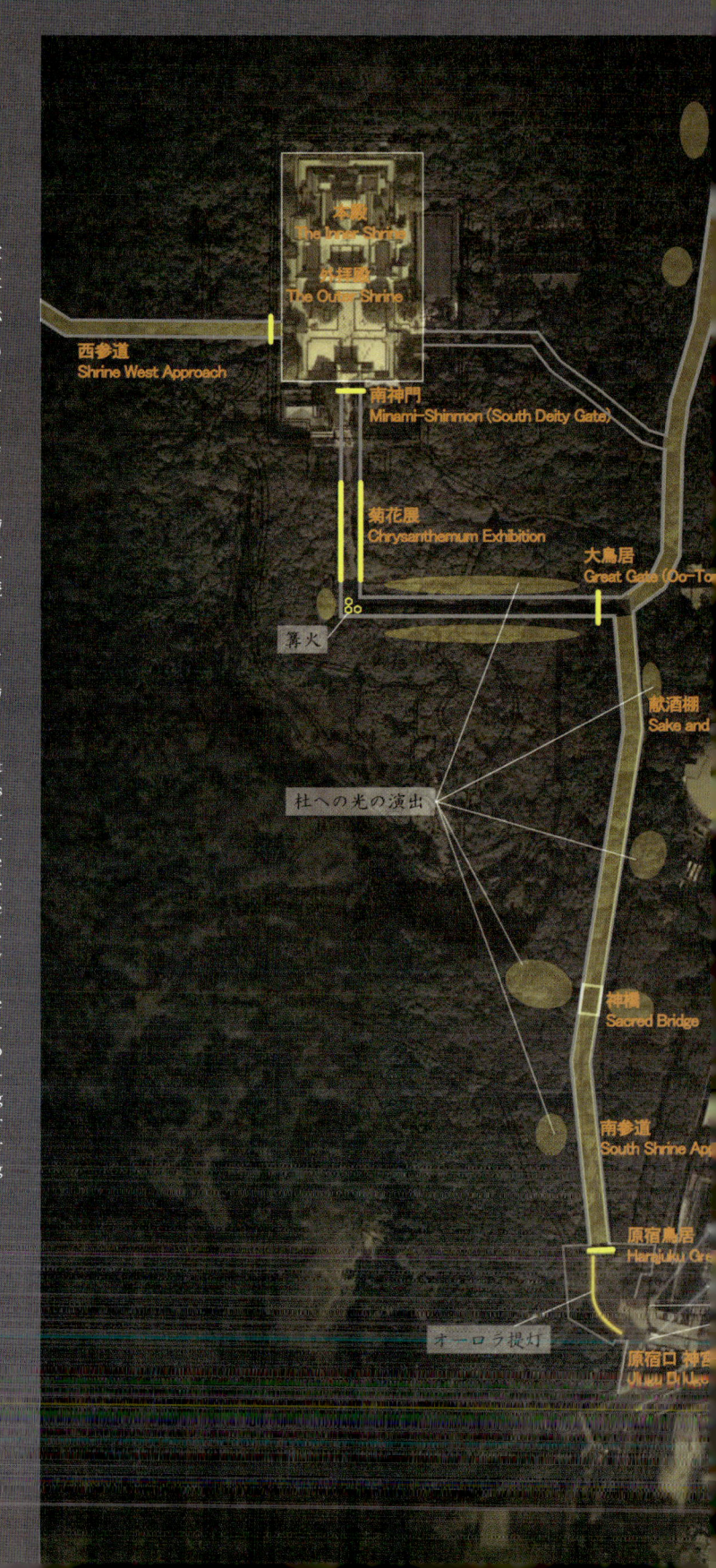

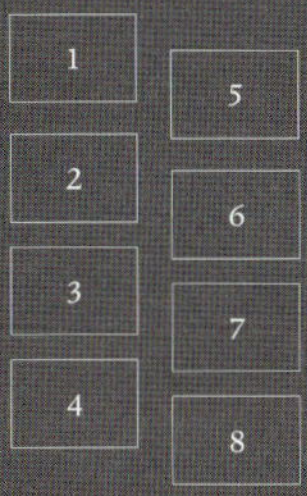

1	5
2	6
3	7
4	8

1　原宿門に設置された光の鳥居
2　LEDと白熱灯の競演、オーロラ行灯
3　杜の中に煙るミステリアスな光
4　建物の中からあかりが溢れる南神門
5　南参道から原宿門を振り返る
6　南参道に敷かれた青い光のじゅうたん
7　大鳥居のライトアップ
8　御社殿の外観

1　A Gate of Light at the Harajuku main gate
2　LEDs and incandescent lamps compete with each other simulating aurora lights
3　Mysterious light in a stand of trees
4　Light glows from inside of the south gate
5　Looking back toward the Harajuku Gate along the south approach
6　A carpet of blue lights along the south approach
7　Illuminating the main gate
8　The main pavilion façade

闇を知る
KNOWING DARKNESS

左｜北参道に覆いかぶさる木立をライトアップしたら、少し明るさ感が出た
中｜北参道はメインの導線から少し外れた位置にあり人通りも少ない参道だ。すれ違う人の表情さえ判別できないほどの暗さが残された
右｜西の広場は広い芝生に覆われているので、500個ほどのキャンドルを配置してあかりの海原を演出した

L: The west plaza is a grassy field. Five-hundred candles were placed here to create a sea of candle flames
C: The north approach is off the main areas where most visitors walk. Not many visitors use this approach and it is thick in darkness; so dark one cannot even recognize the facial features of others walking by
R: We illuminated some of the trees and leaf canopy covering the north approach, creating a sense of brightness

ゼミ生担当の北参道付近には、ゆったりと光の塊が移動する森の景色をプログラム演出した。青色LEDに拡散シートを被せ浮遊する光の聖を出現させた

The design studio students were in charge of the north approach and they designed a program of clusters of light slowly drifting through the forest. Using diffusion sheets over blue LEDs they achieved a kind of suspended spiritual light

闇を知る
KNOWING DARKNESS

面出薫――火と光の記憶　原研哉

面出薫の光に対する感受性を感じたのは、2000年に開催した「RE DESIGN―日常の21世紀」という展覧会で「マッチ」をリ・デザインしてもらった時のことだ。当時の面出薫は「照明探偵団」なるものを組織し、都市の明かりの様相をつぶさに観察しては報告をするという、興味深いデザイン活動を行っていた。その視点に興味を覚え、マッチのリ・デザインをお願いしたのである。

マッチというのは、こすって小さな火を灯す道具である。デザインし直す余地などなさそうに見えるかもしれないが、面出薫の回答は見事だった。地面に落ちた、枯れた小枝を集めて、ちょうどマッチ棒の長さに切りそろえ、それぞれの先端に発火剤を付けるというアイデア。簡単なアイデアに見えるが、そこには実に奥の深い光への洞察がうかがわれる。

生の火に触れる機会から僕らは徐々に遠ざかりつつある。台所の調理は電磁調理器が行うようになったし、ガスを使うとしても、発火や消火の仕組みは機械的である。タバコを吸う習慣は社会から葬り去られようとしているが、これにこっそりと火をつける際も、ほとんどがライターである。喫茶店やレストランでマッチを置いているところもめっきり少なくなった。しかし、それでも時に、僕らはマッチの火を手にすることがある。たとえば、バースデーケーキのロウソクに火を灯す時。食卓にキャンドルをしつらえて、いい気分で食事をしようとする時。そういう時に「生きている火」の特別な存在を、ちら、と心の片隅で反芻しているのだ。

人間と火の関係は実に深く長い。人類を他の動物と分かち、人類たらしめたのは「火を使う」という行為であった。直立歩行によって自由になった両の手の平の中に、差配できる小さな火を携えることから、人の文化は始まったのである。

面出薫のマッチは、自然の木の枝の先に灯された原初の火のメタファである。手の平の中に、原初の火に通じる生きた炎を灯してしてみることで、僕らは火の尊厳、灯りの切実さに、素直な気持ちをむけることができる。「記念日のためのマッチ」と題されていたので、これは特別な日のために火を灯す道具であろう。それは晴れの日のキャンドル点灯か、慎ましさを育む心を寄せ合う儀式か。数万年の人類史の中で灯されてきた営みの深い喜びと充足を、僕らはそこに辿ることができるのである。

小枝の先の火は小さいが、それは地を焼き尽くす強大な火に通じている。生と死、創造と破壊は表裏をなす。使い道を誤ると、火は人類の存亡をも揺るがす地獄の業火と化す。今日、人類は、原子力の火を使いあぐねている。

小枝に灯った小さな光は、そのようなあらゆる火と光の可能性の縮図を、僕らの手の平の中に描き出そうとするものであった。まさにそこに、面出薫の光へのアプローチがあるように思えて、僕は「光のデザイナー」の感受性というものを、克明に理解できたように感じたのである。

原 研哉　*Kenya Hara*

グラフィックデザイナー/1958年岡山県生まれ。1983年日本デザインセンター入社。1991年、社内に独立セクションとして原デザイン研究室（現原デザイン研究所）設立。広告、アートディレクション、展覧会などを手掛ける。武蔵野美術大学基礎デザイン学科教授。日本デザインセンター代表取締役。日本デザインコミッティー理事長。日本グラフィックデザイナー協会副会長。

Graphic designer/Born in Okayama Prefecture in 1958. Joined Nippon Design Center, Inc. in 1983. Established in the Center the Hara Design Laboratory (now Hara Design Institute) as a separate section in 1991. Engaged in advertisement, art direction and exhibitions. Professor, Department of Science of Design, Musashino Art University. Representative Director, Nippon Design Center. Director, Japan Design Committee. Vice Chairperson, Japan Graphic Designers Association Inc (JAGDA)

Kaoru Mende — Memories of Fire and Light

At my request, Kaoru Mende redesigned matches for an exhibition entitled "RE DESIGN—The Quotidian 21st Century," held in 2000. It was then that I became aware of his sensitivity to light. At the time, Mende was engaged in a very interesting design activity; he had organized a group called "Lighting Detectives" that carefully observed and reported on modes of lighting in cities. Intrigued by his viewpoint, I asked him to redesign matches.

A match is a tool that ignites when rubbed to create fire. There may seem to be little room for its redesign, but Mende had a great solution. His idea was to gather fallen twigs, cut them into the same lengths as matchsticks, and cover the tip of each with ignitable material. The idea may seem simple, but it shows real insight into the nature of light.

We have fewer opportunities these days to come into direct contact with fire. We use microwave ovens to cook in the kitchen; even if we use gas, igniting or extinguishing a flame is done mechanically. Society is trying to do away with the smoking of tobacco, and even if it is done surreptitiously, most of the time the cigarette is lit by a lighter. Very few coffee shops or restaurants provide matches now. Nevertheless, we still use matches occasionally, for example, in lighting the candles on a birthday cake or candles set on a dining table to create a pleasant atmosphere. At such times, we briefly ponder the special significance of a live flame.

Humankind has had a long relationship with fire. The use of fire separated us from other animals and made us human. Human culture can be traced back to our being able to carry burning pieces of wood—manageable bits of fire—because walking upright had freed our hands.

Mende's matches were an allusion to primeval fire burning at the end of a branch. Holding in our hands a thing on fire that evokes the primitive fire of humankind helps us see once more the majesty and consequence of fire. His matches were entitled "Matches for a Memorial Day," so they were probably meant as devices for lighting fires on special occasions; for example, to light candles for some formal event or some ceremonial gathering meant to teach respect. We can sense the deep joy and contentment experienced on those countless occasions when lights were lit during the tens of thousands of years of human history.

The flame at the end of the twig is small but linked nonetheless to conflagrations that have scorched the earth. Life and death, creation and destruction are but opposite sides of the same coin. Used in the wrong way, fire can become the flames of hell threatening humankind's existence. Today, humankind has become increasingly tired and wary of using the fire of nuclear power.

The small flame lit at the end of a twig was an attempt to suggest to us, by means of an object at our fingertips, all these different possibilities of fire and light. It seemed to me to symbolize Mende's approach to light. I felt I was able to get a deeper understanding of the sensibility of this designer of light.

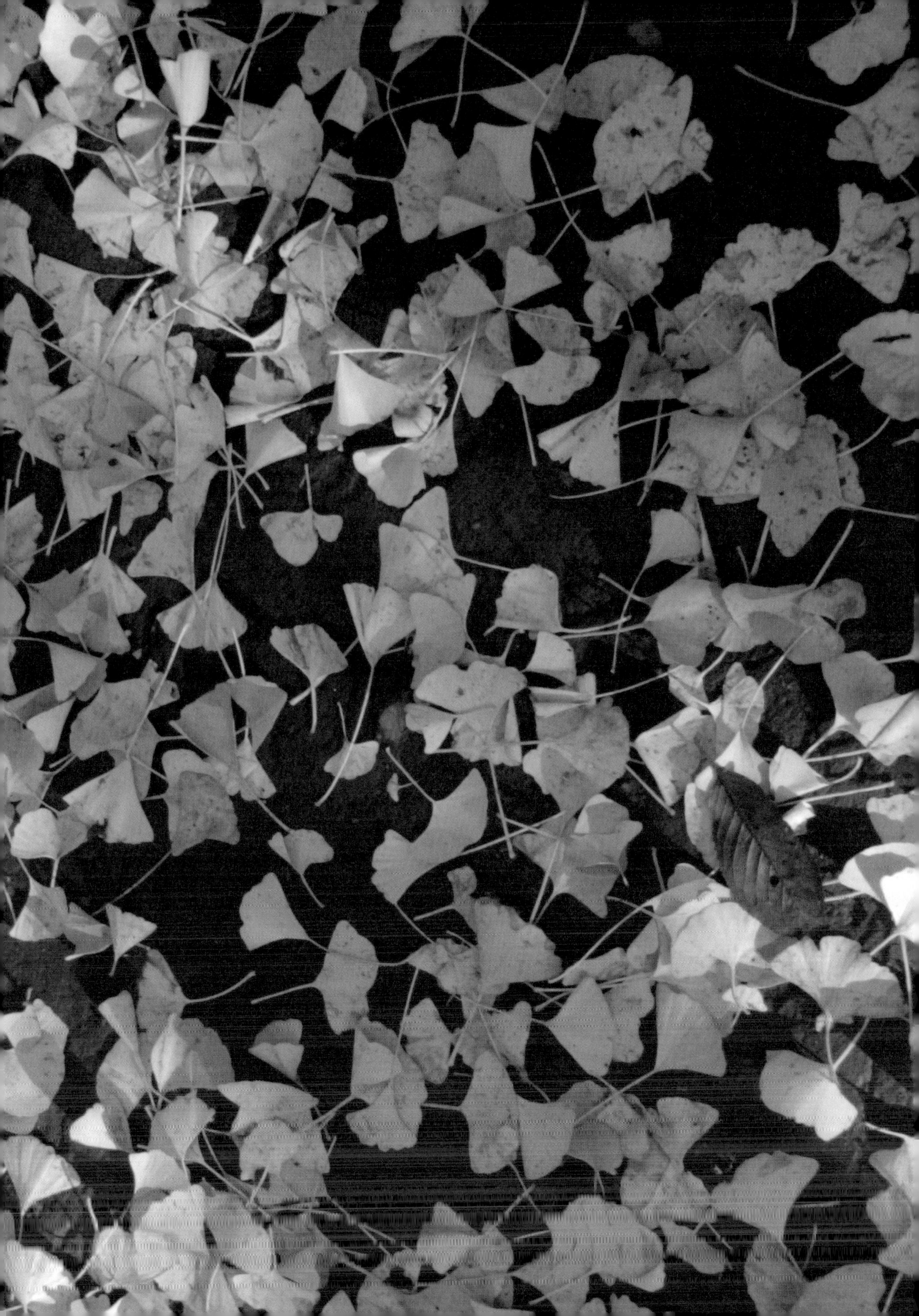

影と遊ぶ
PLAYING WITH SHADOWS

126

影と遊ぶ
Playing with Shadows

この場合、影ではなく陰影という方が正しいのかもしれないが、私は常に光の化身として陰や影をいとおしんでいる。光輝くものの実態より、光によって創作される陰影の様相の方がはるかに深遠であり、心休まる情景を作ることに役立っている。眩しい太陽を見上げずに、地表に降り注ぐ涼しい木漏れ日を楽しむかのごとくだ。木漏れ日は光でなく、揺れ動く影にこそ心打たれる。

実際、影のかたちや対比の強さを凝視すると、その空間にあるすべての光源の様子が推測できる。「さあ今から天井や壁を一切観察せずに、1本の鉛筆を机上に立てて、その鉛筆の影から推測できるこの環境の光を解説せよ」というクイズを学生に出すことがある。机の上に立てられた1本の鉛筆からは、濃い影、薄い影、長い影、短い影、色の着いた影、消え入りそうな影など、たくさんの理由のある影が伸びている。天井の蛍光灯、白熱灯のダウンライト、窓際から忍び込む拡散した自然採光、それらは確実に異なる影を作っている。陰影の上手な作り方、それが照明効果の品質であるといっても過言でない。

影は昔からたくさんの遊びを通して楽しまれてきた。世界中に影にまつわる遊びがありデザインがある。日本風の陰影はぼんやりと拡散して、時にグラデーションが効いたものも多いが、西洋の陰影は光と影とのコントラストが強いものが主流だ。日本人がイエス・ノーのはっきりしない態度を得意とするのに対して、欧米人は二元論に基づく論理を得意とする。室内で使うキャンドルの光でさえ、灯明の直射光を和紙で包み込み柔らかな陰影を意図した日本人とは異なり、欧米ではその揺れる点光源が明快な陰影の対比を演出する。柔らかい影と硬い影。まさしく気候風土や文化、宗教の違いが作る陰影のバリエーションだ。

僅かなキャンドルの光を使いながら陰影の重要性を学習する、という課題をいくつか出題した。2003年より10年間継続したキャンドルナイト＠Omotesando という学外プロジェクトは、まさに僅かなキャンドルの光を頼りに、繊細な陰影を表現するための絶好の機会となった。「あかりを消してスローな夜を」というスローガンに支えられた表参道を舞台にしたパフォーマンスだったが、街のあかりの量だけは期待通りに少なくなっては行かなかった。

Instead of shadow, it may be more correct in this instance to say shade and shadow. I always value shade and shadow as incarnations of light. Shade and shadow created by light have far greater depth than something radiant and help to create a restful scene. Gazing at dappled light falling on the ground is far more enjoyable than looking up at the dazzling sun. What is impressive about dappled light is not the light itself but the trembling shadows accompanying the light.

In fact, if we look at the shape or the strength of contrast of a shadow, we can deduce the condition of all light sources in that space. I sometimes give students this quiz: Stand a pencil on your desk, and without looking at the ceiling or the walls, deduce the condition of light in this environment from the shadows cast by the pencil. Many different shadows stretch from the pencil: dark shadows, light shadows, long shadows, short shadows, colored shadows and faint shadows. The fluorescent lamps and the incandescent down lights on the ceiling and the dispersed natural light sneaking in from windows—these create definitely different shadows. It would not be an exaggeration to say that the skill with which shade and shadow are created determines the quality of any lighting effect.

Since long ago, there have been many ways of playing with shadows. There are games and designs using shadows found throughout the world. Japanese-style shades and shadows tend to be vague, dispersed and characterized by subtle gradation, while in the West there is often a sharp contrast between light and shadow. This may reflect a cultural difference. The Japanese prefer to adopt an ambiguous attitude instead of being for or against something; meanwhile, Westerners prefer a logic based on dualism. There is a difference even with interior candlelight; the Japanese will wrap washi around the light to soften the contrast between light and shadow, while in the West, the flickering light source is allowed to produce a clear contrast between light and shadow. Delicate shadows and hard shadows--these variations in shadow indeed reflect differences in climate, culture and religion.

I have given students a number of assignments involving the faint light from a candle so that they can learn the importance of shade and shadow. An extramural project called "Candlenight@Omotesando" that I have organized for ten consecutive years starting in 2003 provides a perfect opportunity to express delicate shades and shadows using the faint lights of candles. This is a performance staged in Omotesando that urges people to "turn off lights and slow down the night," but lights in the district have not decreased as we had hoped.

キャンドルナイト─東京・表参道
Candle Night in Omotesando, Tokyo

　私が大学に赴任した翌年から「100万人のキャンドルナイト」という環境キャンペーン運動が日本全国で始まった。文化人類学者の竹村真一さんからの要請で、私はゼミ生を引き連れて2003年冬至から10年間に渡りこれに継続的に参加した。

　私は1本のローソクが創りだす陰影の妙を学生に知ってほしかった。2003〜2005年は原宿キャットストリートで、その後2006〜2012年は表参道ケヤキ道（表参道欅通り商店街）に場を移して、10年間に渡り地元の方々と一緒のワークショップを行った。都内の美術系大学を中心に多くの大学生や専門学校生が参画し、当日の参加学生は最大で600名あまりにも達するイベントに拡大したので、2012年をもって10年間の継続イベントを終了した。

www.candle-night.org

"Candle night for million people", an environmental campaign, was started throughout Japan one year after I took a post in the university. At the request of Mr. Shin-ichi Takemura, an anthropologist, I continuously participated in the campaign with the students of my seminar for nearly ten years since midwinter of 2003.

　I wanted my students to know the subtle shading created by a single candle. We conducted workshops with local residents for ten years, on the Harajuku Cat Street in 2003 through 2005 and then on the Omotesando Avenue lined with zelkova trees (in a shopping area along the Keyaki-dori Avenue) in 2006 through 2012. Students mainly from fine art universities in Tokyo and numerous other universities and vocational schools participated in the campaign. As the maximum daily participation reached 600 students, the ten-year event was discontinued in 2012.

キャンドルナイト10年間の系譜
Genealogy of Candle Night for 10 years

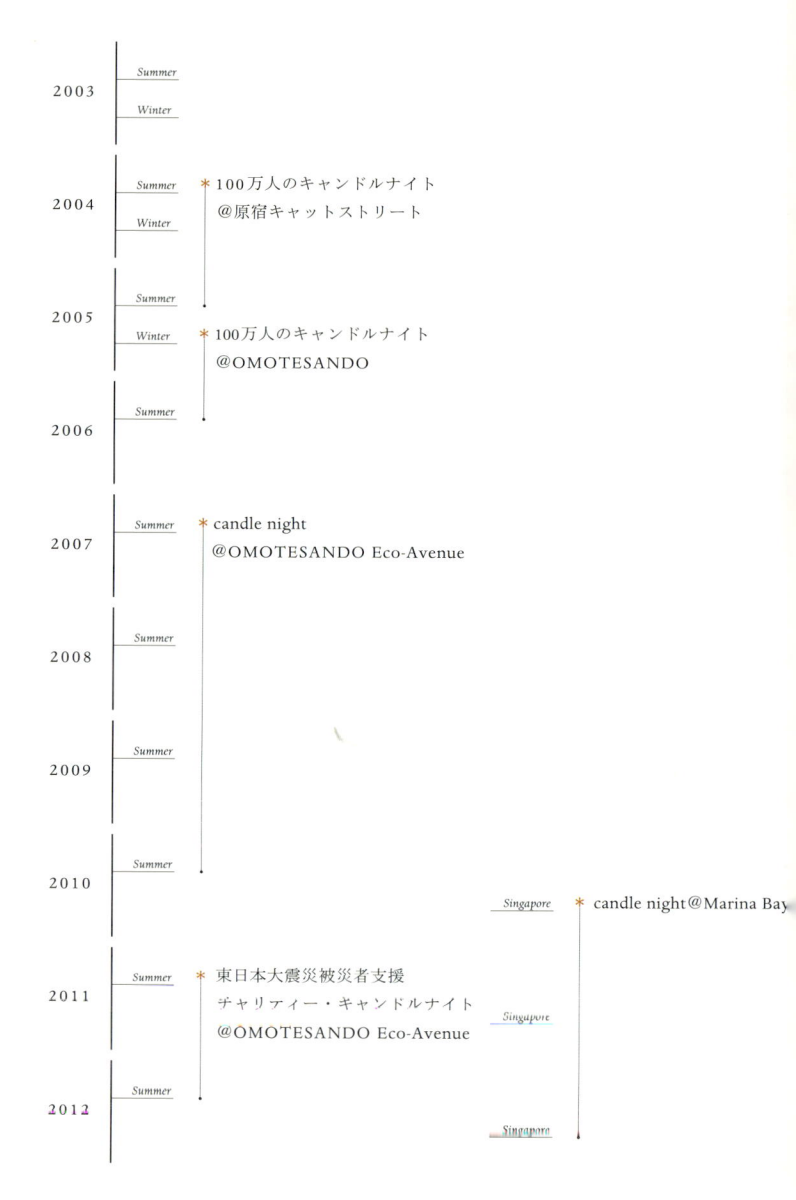

年	季節	イベント
2003	Summer	
	Winter	
2004	Summer	＊100万人のキャンドルナイト @原宿キャットストリート
	Winter	
2005	Summer	
	Winter	＊100万人のキャンドルナイト @OMOTESANDO
2006	Summer	
2007	Summer	＊candle night @OMOTESANDO Eco-Avenue
2008	Summer	
2009	Summer	
2010	Summer	Singapore ＊candle night @Marina Bay
2011	Summer	東日本大震災被災者支援 チャリティー・キャンドルナイト @OMOTESANDO Eco-Avenue / Singapore
2012	Summer	Singapore

2010年夏至、佐藤卓さんが声を振り絞ってカウントダウンを行う
Taku Satoh leads the countdown at the 2010 Summer solstice

竹村真一
Shinichi Takemura

発起人＋監修者の
竹村真一さんは皆勤賞

*Founder + Supervising Editor,
Shinichi Takemura provided
annually perfect attendance*

浜野安宏
Hamano Yasuhiro

初代の監修者・
浜野安宏さん（2004年）

*The first supervising editor,
Yasuhiro Hamano, 2004*

深澤直人
Naoto Fukasawa

深澤直人さんも監修者として
講評にも参加してくれた

*Naoto Fukasawa gave the student
feedback as a supervising editor*

佐藤 卓
Taku Satoh

監修者の佐藤卓さんは
学生たちに大人気

*Supervising Editor, Taku Satoh
is popular among students*

キャンドルナイト —— 4つのパフォーマンス
Candle Night —— Four main performances

キャンドル・インスタレーション
Candle Installations

日本看護協会、表参道ヒルズ、交番横まちかど広場、八千代銀行まえ、オリエンタルバザー、キャットストリート入口広場、などにご協力いただき、キャンドルによる環境アートづくりを行った。

Various tenants including, the Japanese Nursing Association, Omotesando Hills, Machikado Square, Yachiyo Bank, Oriental Bazaar, and Cat Street Square, lend space for environmentally-minded candle art installations.

キャンドル・カフェ
Candle Café

欅道とその周辺に位置するお洒落なカフェに協力を依頼した。同日の夜2時間だけを学生たちが提案するキャンドルのあかりで演出する、というものだ。20店舗ほどの協賛をいただいた。

In cooperation with several trendy cafes in the Omotesando area, for two hours on Candle Night, candles designed by college students are placed on tables. About 20 cafes are included in this network.

オリジナル行灯
Original Lanterns

学生たち個人が提案するキャンドルを用いた携帯用の行灯。歩道などの公共空間には行灯を設置することが許可されないので、制作された行灯は当日に学生たちが携帯して欅道を巡回展示した。

College students are encouraged to propose and design their own portable lanterns for this one-night event. Hanging lanterns along the sidewalks of Omotesando is prohibited; however students are allowed to carry their own original lanterns up and down the tree-line street as they move between candle installations.

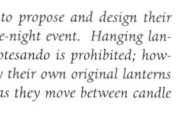

子供ワークショップ+パレード
Children's Workshop + Parade

神宮前小学校の子供たちと一緒にキャンドルによる行灯づくりを行った。当日の夕方には父兄と一緒に子供たちの行灯パレードも行われ、キャンドルナイトの風物詩になって行った。

College students help children of Jingumae Elementary School make lanterns for a night parade. Along with parents and family members, the children walk along a parade route with their lanterns. The children and their lanterns have become a charming feature of Candle Night.

キャンドルナイトは地元の方々の指導と協力をいただき、商店に2時間の消灯や減灯をお願いしたり、神宮前小学校の子供たちとのあかりワークショップも行った。毎年のイベントは、キャンドル・インスタレーション、キャンドル・カフェ、オリジナル行灯、子供ワークショップ＋パレード、の4つのパフォーマンスで構成された。

Candle Night is an event held under the direction and with the support of local people. On the night of the event, shops dim lights or go completely dark for two hours and, among other scheduled activities, local students at Jingumae Elementary School participate in an Akari Workshop. This annual event is composed of four main performances: Candle Installations, Candle Cafes, Original Lanterns, and Children's Workshop + Parade.

キャンドルナイト@表参道 開催概要
Candle Night@Omotesando Event Brief

表参道という現場をもって学習する。10年間のキャンドルナイトには様々な取り組みがなされたが、何よりも表参道という目立った現場の発する力は偉大だった。表参道だからやりがいがあったし、1日のみのイベントだが多くの通行人が批評してくれた。デザインの学習には、大学のキャンパスから離れることがより重要なこともある。

Learning from Omotesando. Over the ten years that Candle Night was held in Omotesando, we tried various things to shake up the yearly event. But the greatest factor in the success of the event was the energy emanating from the area itself. Because the location was Omotesando the challenge was worthwhile. It is only a one-night event, but those unexpecting shoppers who just happen to walk by made comments and inquiries. As students of design, getting away from the university campus ended up being a very important lesson.

明治神宮
Meiji Shrine

JR 原宿駅
Harajuku Station

凡例
a 開催日
b 監修
c 参加人数
d イベント拠点数
e 参加校

Candle Cafe
東京アパートメントカフェ@原宿
Tokyo Apartment Cafe @ Harajuku

Candle Cafe
京橋千疋屋表参道原宿店
kyobashi sembikiya omotesando harajuku

Installation
八千代銀行
Yachiyo Bank

オリジナル行灯
Original ANDON
展示エリア
Event area

Candle Cafe
Bakery Cafe 426 Omotesando

エコファームカフェ 632
ECO FARM CAFE 632

大型ビジョン
Large-size screen

生活の木
Tree of Life

Candle Cafe
montoak

Candle Cafe
ロカンダ・エッフェ
LOCANDA.F.Q

Candle Cafe
カルミネ表参道スタンド
Carmine Omotesando Stand

Candle Cafe
ファーマーズカフェ
Farmer's Cafe

Installation
キャットストリート入口
Entrance of Cat Street

Candle Cafe
カフェアノ
cafe ano

2003 winter
a 2003.12.22
b 面出薫・竹村真一
　浜野安宏
c 50 人
d 6 箇所
e 武蔵野美術大学
　多摩美術大学

2004 summer
a 2004.06.20
b 面出薫・竹村真一
　深澤直人
c 69 人
d 9 箇所
e 東京理科大学
　明治大学
　慶應義塾大学
　日本大学
　東京藝術大学
　武蔵野美術大学
　中央大学
　多摩美術大学

2004 winter
a 2004.12.21
b 面出薫・竹村真一
　深澤直人・佐藤卓
c 75 人
d 30 箇所
e 東京理科大学
　明治大学
　慶應義塾大学
　日本大学
　千葉大学
　武蔵野美術大学
　東京大学
　多摩美術大学

2005 summer
a 2005.06.19
b 面出薫・竹村真一
c 97 人
d 6 箇所
e 東京理科大学
　明治大学
　慶應義塾大学
　東京工業大学
　東海大学
　武蔵野美術大学
　日本女子大学
　多摩美術大学
　京都造形芸術大学

2005 winter
a 2005.12.22
b 面出薫・竹村真一
c 70 人
d 3 箇所
e 武蔵野美術大学
　多摩美術大学
　東京理科大学
　慶應義塾大学
　東京工芸大学
　東京造形大学
　東海大学
　日本大学

2006 summer
a 2006.06.21
b 面出薫・竹村真一
c 61 人
d 13 箇所
e 武蔵野美術大学
　多摩美術大学
　桑沢デザイン研究所
　女子美術大学
　東京理科大学
　慶應義塾大学

2007 summer
a 2007.06.22
b 面出薫・竹村真一
　深澤直人
c 300 人
d 24 箇所
e 武蔵野美術大学
　多摩美術大学
　桑沢デザイン研究所
　女子美術大学
　芝浦工業大学
　東京造形大学
　東北芸術工科大学
　千葉大学
　慶應義塾大学
　東海大学
　東洋英和女学院大学

2008 summer
a 2008.06.21
b 面出薫・竹村真一
　佐藤卓
c 300 人
d 33 箇所
e 武蔵野美術大学
　多摩美術大学
　桑沢デザイン研究所
　女子美術大学
　芝浦工業大学
　東京デザイン専門学校
　東京造形大学
　法政大学
　関東学院大学
　慶應義塾大学

	2009 summer	2010 summer	2011 summer	2012 summer
a	2009.06.19	2010.06.19	2011.06.17	2012.6.22
b	面出薫・竹村真一・佐藤卓	面出薫・竹村真一・佐藤卓	面出薫・竹村真一・佐藤卓	面出薫・竹村真一・佐藤卓
c	482 人	611 人	452 人	345 人
d	35 箇所	27 箇所	19 箇所	12 箇所
e	武蔵野美術大学 多摩美術大学 桑沢デザイン研究所 女子美術大学 芝浦工業大学 東京造形大学 東京デザイン専門学校 東洋美術学校 関東学院大学 慶應義塾大学 デジタルハリウッド大学	武蔵野美術大学 多摩美術大学 桑沢デザイン研究所 女子美術大学 芝浦工業大学 東京デザイン専門学校 東京造形大学 東洋美術学校	武蔵野美術大学 多摩美術大学 桑沢デザイン研究所 女子美術大学 芝浦工業大学 東京デザイン専門学校 東京造形大学	武蔵野美術大学 多摩美術大学 桑沢デザイン研究所 女子美術大学 芝浦工業大学 東京デザイン専門学校 東京造形大学

Candle Cafe
ピアザ エコファーム カフェ
PIAZZA ECO-FARM CAFE

Candle Cafe
ベーカリーカフェ 426 表参道
Bakery café 426 Omotesando

Installation
キャットストリート入口
Entrance of Cat Street

Installation
表参道まちかど庭園
Omotesando Machikado Teien

神宮前小学校
Jingumae Elementary School

Installation
表参道ヒルズ キャナル
Canal at Omotesando Hills

子供たちの
キャンドルパレード・ルート
Children's Candle Parade

表参道
新潟館ネスパス
Niigatakan N'ESPACE

Installation
オリエンタルバザー
Oriental Bazaar

Candle Cafe
イルピノーロ カフェ
IL PINOLO CAFFE

Installation
日本看護協会(大階段・広場・ピロティ)
Japanese Nursing Association
(Grand Staircase, Plaza, Piloti)

Candle Cafe
表参道バンブー
bamboo at Omotesando

Installation
表参道ヒルズ 貫通通路
Passing path at Omotesando Hills

Candle Cafe
表参道ヒルズ内 Shops in Omotesando Hills
洋食ミヤシタ YOSHOKU MIYASHITA
トラヤ カフェ TORAYA CAFE
Trattoria & Pizzeria Zazza
表参道茶寮 Omotesando Saryo
やさい屋めい YASAIYA·Mei
ビスティーズ BISTY'S
R Style by 両口屋是清 R style -Fine Japanese Confectioneries
Gelateria BAR naturalBeat
キュウブ ゼン CUBE ZEN
DEL REY Cafe & Chocolatier
ミスト MIST
はせがわ酒店 Hasegawa Sake Shop
ポワヴリエ POIVRIER
クルックスリー kurkku 3

Installation
表参道ヒルズ 水景
Water feature at Omotesando Hills

Candle Cafe
ナル カフェ
NARU Cafe

Candle Cafe
アニヴェルセル カフェ
ANNIVERSARE CAFÉ

Candle Cafe
ブラウンライスカフェ
BROWN RICE CAFE & DELI

シュウウエムラ
shu uemura

Goal / Start

0m 50m 100m

監修者、佐藤卓さんと面出薫によるインスタレーション案の中間チェック風景
Supervising Editors, Taku Satoh and Kaoru Mende attend a midterm review of installation proposals

Six months of preparation

Initially, Candle Night was held twice a year on the summer and winter solstice. Soon the scale of the event grew and turning off the lights during the Christmas holiday and commercial season was too much. Now it is an annual event on the summer solstice in June. Also, to hold this local event there is a mandatory protocol and several required formalities to manage, which takes about six months. A few examples are: a kick-off presentation to the administrative board of the local merchant's association, Keyakikai, formal requests to local tenants to dim or turn off lights for the event, apply for event permits with the local fire department and police station, meeting with the local Jingumae Elementary School, appeal for co-sponsors, press releases, design publicity flyers and such…the list goes on.

Student volunteers from local universities form the planning committee to divide up all of the work and responsibilities. These student volunteers hold regular briefings with the supervising editors to check on various matters and exchange ideas. Also, they visit all local tenants to brief them on the event, an essential part of the preparation process. On the next page is a typical record of the six month process.

6か月に及ぶ準備プロセス

初期的に私たちのキャンドルナイトは夏至と冬至の年2回行っていたが、暫くするとその規模も大きくなり、冬至のクリスマス商戦では消灯が難しくなるなどして、夏至（6月）のみに行うことにした。というのも、地元でのイベント開催には様々な手続きが義務付けられ、それをこなす日程管理にも6か月ほど要するようになっていったからである。

　欅会商店会理事会へのイベント企画プレゼ、地元商店への消灯・減灯協力要請、消防署と警察へのイベント申請、神宮前小学校との協議、協賛のお願い、プレス・広報物の作成などなど…。

　各大学有志で実行委員会を作り責任と作業を分担する。監修者との定期的なチェックと意見交換、地元へのあいさつ回りなども準備のプロセスに不可欠だ。次ページに示すのは、その典型的な6か月プロセスの記録メモである。

表参道ヒルズの水景インスタレーション準備
Preparing the water feature at Omotesando Hills

オリジナル行灯の投票用コンテスト・ボックスの制作
Creating the ballot box for the Original Lantern Contest

candlenight@OMOTESANDO – Eco Avenue 2011夏至 スケジュール
candlenight@OMOTESANDO実行委員会

- 実行委員担当決定
- 希望場所の確認
- 担当仮決定
- **担当決定**
- 希望場所が重なった場所はコンペ形式にて選出
- 制作確認
- 制作確認

項目	担当	1月		2月		3月
実行委員会		*1/14 19:00〜21:00 キックオフmtg	①1/28 19:00〜20:00 実行委員mtg		*②2/22 18:00〜19:00 実行委員mtg	*③3/11 18 実行委員m
総合管理	藤井+池田+井本+東		仕事内容説明		①2/22 19:00〜21:00 デザインチームmtg	デザインチー
実行委員メンバー決定						
担当決定						
キャンドルインスタレーション	池田/学生①		希望場所 確認		アイデア発表	試作
場所選定・依頼						
材料調達						
子供たちのキャンドルパレード	東/学生②				企画概要説明	3月中
神小とのmtg・ワークショップ ←日程至急check						
企画						
作品制作・材料調達						アイデア発表	試作品
キャンドルカフェネットワーク	藤井/学生③		希望場所 確認		アイデア発表	試
場所選定・依頼 HG					3月上旬関係各所との調整	
作品制作・材料調達						
オリジナル行灯	藤井/学生④		希望確認		アイデア発表	試作
場所選定・依頼 IMO						
作品制作・材料調達						
樽会	藤井+東					
協賛金呼びかけ						
企画書						
事前挨拶・主旨説明				樽会挨拶	理事会説明	SHOP連絡会説明
広報	池田+井本/学生⑤					
HP作成+更新		更新			素案確認	制作
ポスター+フライヤー 作成・配布 ←承承			デザイン		素案確認 2月中に〜リクインでPDで	制作
グッズ作成(マッチ等) 中止			デザイン			
大型ビジョン/ヒルズビジョン映像作成 中止					コンテ作成	素案確認
TV映像作成 中止					コンテ作成	素案確認
プレス発表	東	プレス先の検討				プレスへの
記録	池田					
写真・VTR IMO						
その他	−					
カメヤマローソクとの調整					連絡・調整	3/10 17:00~2
警察消防関係	藤井+東					
表参道ヒルズ/生活の木					連絡・調整	
看護協会/新潟ネスパス	池田+東					連絡・調整
減灯・消灯のお願い	IMO 池田+学生⑥					連絡・調整

2011年2月10日 運営

- キャンドル希望個数提出 HIG, NOK
- 広報ツール紹介

- 制作最終確認 ← NOK
- キャンドル配布
- 広報ツール配布

当日のスケジュール取り
- CD映像お披露目
- 当日の確認

4月	5月	6月	7月
*④4/15 18:00〜19:00 実行委員mtg		*⑤5/26 19:00〜21:00 実行委員mtg	*⑥6/10 19:00〜21:00 実行委員mtg	*7/8 打上
*③4/15 19:00〜21:00 デザインチームmtg	*⑤5/20 19:00〜21:00 デザインチームmtg			

演出+佐藤暑の

	試作品確認/企画書提出	最終確認		
		5月初旬 現地実験 本制作		
会議提	試作品確認/企画書提出	学校への説明/内容承認/参加者募集 最終確認		神宮前小学校でのワークショップ 6/14あたり
		本制作		
出	試作品確認/企画書提出	最終確認		
		本制作	18:00-21:00 QLIA	
			不ジュール確認会 6/1 12% 6/2 29%	
出	試作品確認/企画書提出	最終確認	6/1-2 演出さんチェック	
	5/16 スタッフチェック	メインスタッフ本制作	5/24 スタッフチェック 当日スタッフ本制作	予備日（6月18日/土） 当日（6月17日/金） 夏至（6月22日/水）
	企画書まとめ 4/中旬欅会へ個別の企画書提出			
	各チーム 企画書提出		欅会事前説明	
高校ジュール (作).		更新		
	発表 4月中旬〜後半入稿	配布		
	発表 4月中旬〜後半入稿	配布		習日にも戻ってきて該当する
	中間チェック	5/13 17:00-19:00 演出さんチェック 発表	5/26 19:00-20:00 演出さんチェック データ変換放送スタート	
	中間チェック	発表		
	取材・広報			
			説明会	まとめ映像着手
キャンドル数連絡	4/中旬 キャンドル個数決 連絡・調整			
			書類作成/提出	

企画書作り、イベント内容に変更。

オリジナル行灯の制作

美大生が多く参加するので、何といってもオリジナリティの高い自作のあかりデザインに興味が集中する。誰もが私ならではのユニークな行灯をデザインして皆に見てもらいたいという気持ちだ。監修者は光り燃えやすいような構造でないことを強要する。下げ提灯のようなもの、布くピニールなどの素材に凝ったもの、風を強く意識したもの、大きさを誇るものなど様々だ。何といっても火の点いたローソクを

使うだけで不自由さが付きまとう。一度だけローソクと一緒に使うことを条件にLEDの使用を認めたことがあるが、あまり好評ではなかった。LEDを使うとあかりのデザインを使ったスローな夜にはならず面白い照明デザインに向かってしまったからだ。しかりローソクの自由さに辛みにかなわない。何十年も前から私たちはその不便さの中に便利さを感じ取っているのだから。

Original lantern design

Many of the Candlenight volunteers are art university students, so needless to say the level of originality and interest is max for lantern design. Everyone wants to design something unique and representative of themselves to showcase at the event. We, as supervisors, first, stress only that designs do not easily catch on fire!

Some lanterns resemble handbags; others use fancy cloth or plastics. Some play with shadows and silhouettes and others boast size, but all are different and unique. However, the inconvenience of a lit candle plagues the amateur designers every year.

One year, LEDs were allowed, if used in conjunction with a candle, but it turned out to be a bit controversial. Using LEDs did lead to some interesting lighting designs, but the idea conflicted with "taking it slow," as the event slogan goes. So, however inconvenient, the candle is the object of the lesson and this is an opportunity to learn from it as we did thousands of years ago, every dark night.

キャンドル・インスタレーション
Candle installation

環境の中に暗闇を見つけて、そこにキャンドルライトを操作することで新鮮な光環境を創りだす、それがインスタレーションのテーマだ。

もちろん光源は電球やLEDではなく不自由なキャンドル。風にも弱いし雨が降っても困ってしまう。しかしまた、それだからこそダイナミックなキャンドルならではの環境アートが出現した。

Find a patch of darkness in the urban environment, add candles, and create a fresh lightscape. This is the theme for candle installations.

Of course, the source of light for candle installations is not LEDs or light bulbs, but the inconvenience of a lit candle. The candle is sensitive to wind and rain is the ultimate enemy. However, because of these challenges, art in the urban environment is possible with the use of the dynamic candle.

看護協会の大階段に出現したあかりの大樹。下から上からの視点場を計算した
A huge tree of light on the steps of the Nursing Association's property. The design is calculated looking up or down the steps

オリエンタルバザー前のインスタレーション
通行人が影絵を演出。幕の内側に半透明のスクリーンを吊りキャンドルを仕込み
Shadow installation in front of The Oriental Bazaar. Candles are set behind a translucent curtain for a candle installation with shadow play

子供たちが制作した行灯をアルミホイルを敷き詰めた広場に展示した
Lanterns made by children are arranged on aluminum foil in the plaza

湾曲した鏡面アルミ板を階段状に並べてあかりの増量を行った
*Curved aluminum panels are arranged on stair steps
and the amount of candlelight is doubled*

オリエンタルバザーは毎年のインスタレーションに店先を提供してくれた
The Oriental Bazaar lends its storefront to the event every year

ジャングルジム全体に赤い和紙を張って巨大行灯に見立てた
*Japanese paper, washi, is taped to the entire surface
of a jungle gym creating a gigantic lantern*

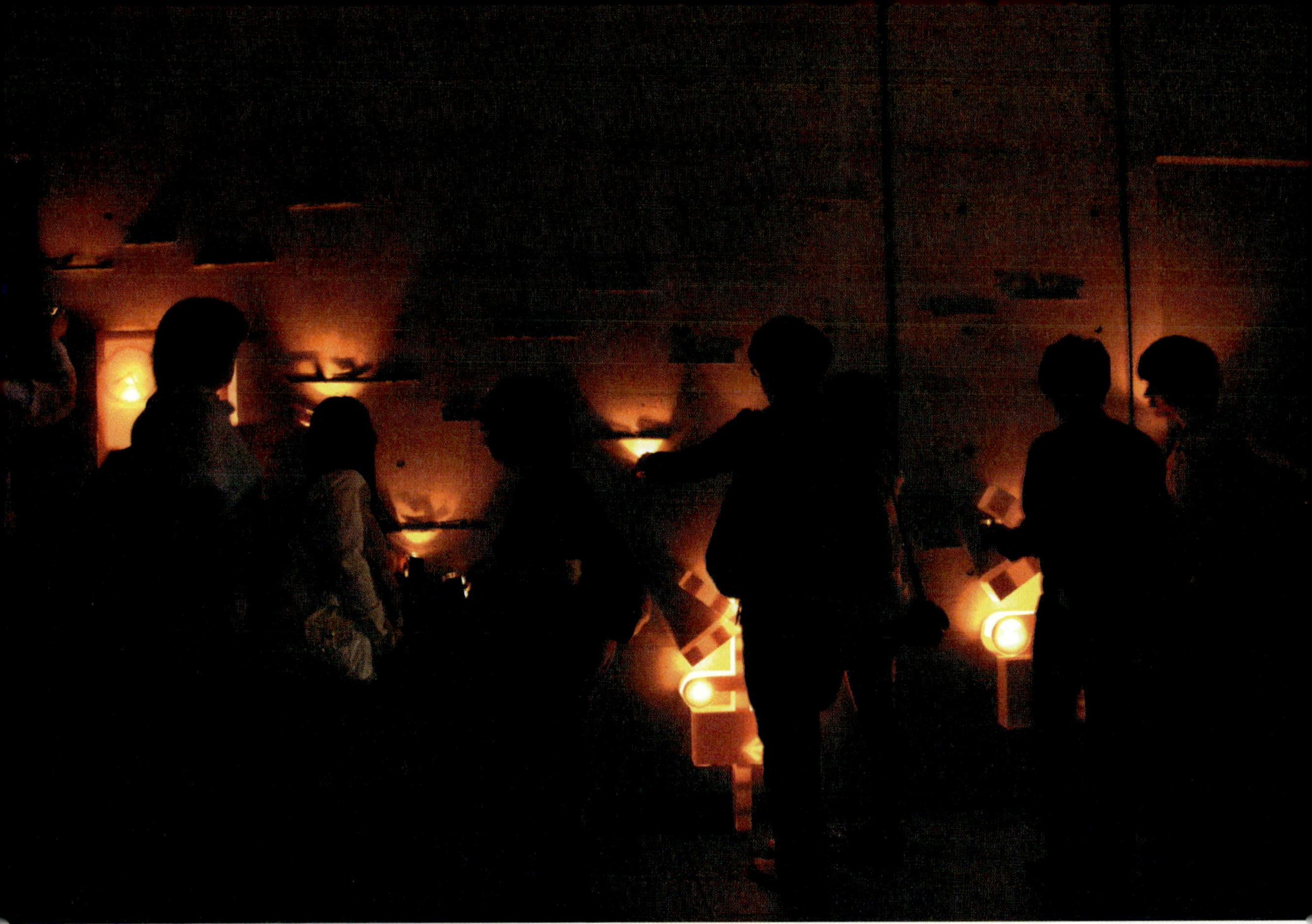

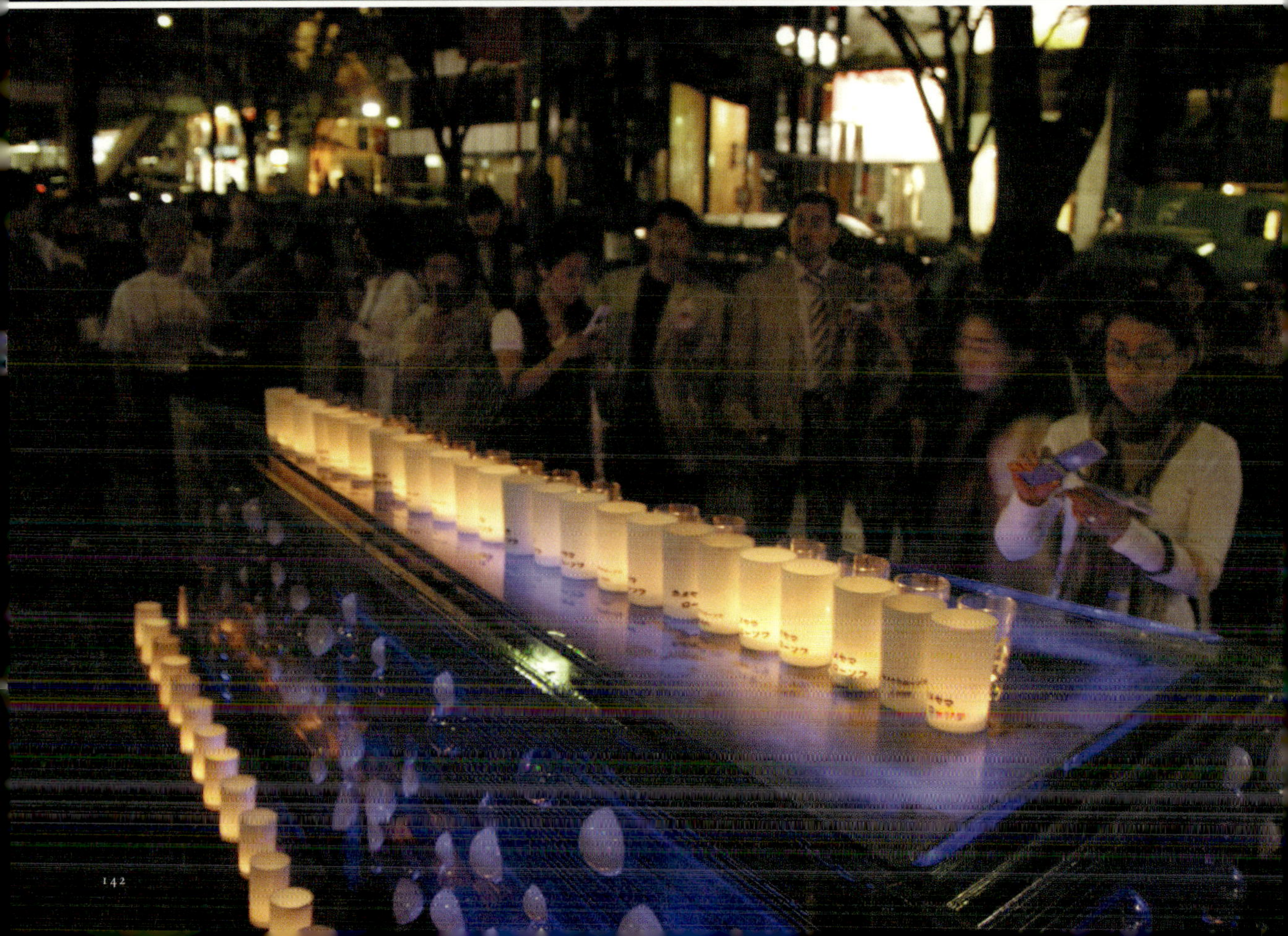

1 表参道ヒルズの貫通通路に作られた
  キャンドル・オノマトペ
2 表参道ヒルズの東端に位置する
  三角形の水景広場
3 キャットストリートの入り口付近には
  流れる川のオマージュが出現した
4 深澤直人さんのオフィスに通じる
  屋外階段。深澤事務所が
  デザイン参加してくれた
5 街角庭園にデザインされた竹あかり
6 白いビニール傘を組み合わせて
  大きな傘行灯が展示された

1 Onomatopoeia candle installation
  in an Omotesando Hills corridor
2 The triangle water feature
  on the east edge of Omotesando Hills
3 A homage to a flowing river
  at the intersection to Cat Street
4 The external stairs
  to Naoto Fukasawa's office, designed
  by the NAOTO FUKASAWA DESIGN
5 Bamboo light in the public garden
6 A large umbrella lantern made from several
  smaller white vinyl umbrellas

1	3	5
2	4	6

144

グッズをつくる
Designing Candle Night publicity materials

キャンドルナイト@Omotesandoの実行委員会ではグラフィック＋広報担当が活躍した。費用が潤沢でないので、大判のポスターなどはできなかったが、毎年工夫を凝らしたデザインで各種のキャンドルナイト・グッズを作成した。その中心となるのがイベント内容を告知するためのハンディなフライヤー。毎年のテーマを活かしたユニークなデザインを行った。ティーキャンドルとマッチをセットにして無料配布を行った年もあった。2011年からは東日本災害復興支援のチャリティーキャンドルなども加わった。

The graphic and publicity chair of the Candle Night@Omotesando Planning Committee was a big success. Even though the budget was far from abundant and luxuries like large publicity posters were impossible, they designed original Candle Night goods every year. But the main event flyer was particularly handy and cleverly depicted the yearly theme. Some years we also handed out free tea candles and matches along with the flyers. In 2011, charity candles were also added to provide aid for the Eastern Japan Disaster and Recovery.

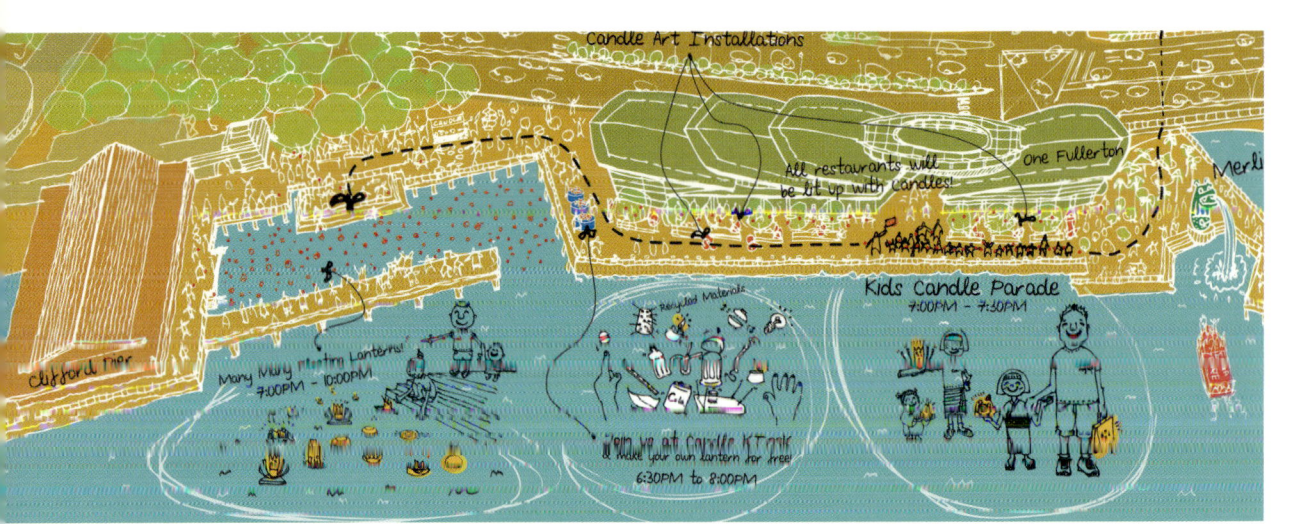

Candle Art Installations

All restaurants will be lit up with candles

One Fullerton

Merli[on]

Clifford Pier

Mary Mary Floating Lanterns!
7:00PM - 10:00PM

Recycled Materials

Jelly up at Candle Kiosk
Make your own lantern for free!
6:30PM to 8:00PM

Kids Candle Parade
7:00PM - 7:30PM

キャンドルナイト―シンガポール
Candle Night in Singapore

2010年10月15日にシンガポール照明探偵団が主催してキャンドルナイト@Marina Bay, Singaporeというイベントが観光名所のマーライオン・パークを中心に行われた。面出ゼミもシンガポールのラサール美術大学などと一緒に積極的に協力参加した。

東京で2003年に始まったキャンドルナイト・イベントを参考にして、URA（Urban Redevelopment Authority）やワンフューラートン・マリーナベイの協力を得て開始したこのイベントは2012年の9月で第3回を数える風物詩になってきた。

On October 15, 2010 near the famous tourist location Merlion Park, the Singapore Lighting Detectives hosted the first Candle Night @ Marina Bay. With the Urban Redevelopment Authority (URA) and One Fullerton Marina Bay as co-sponsors, we used our experiences from Candle Night, Tokyo, which started in 2003, to host the city's third Candle Night in September 2012. Candle Night @ Marina Bay is well on its way to becoming an annual event. Students from the Mende studio at Musashino Art U. and local students from the Lasalle College of the Arts also worked together to help make the event possible.

In Tokyo

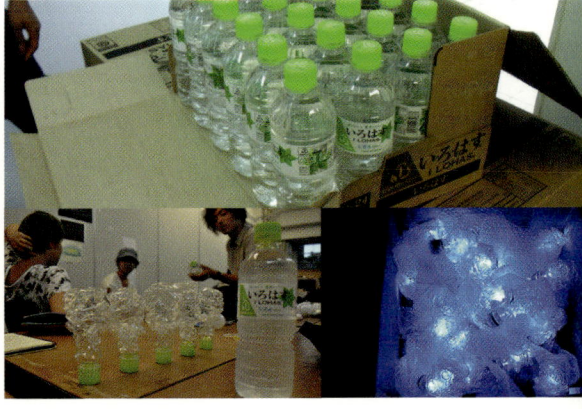

From Tokyo to Singapore

In Singapore

写真左上から
武蔵美ゼミ室での準備風景
ペットボトルを空にしなければならない
「いろはす」の協賛を得て300本の青い灯篭を作った
武蔵美の中庭での制作風景
持ち込んだ制作物と共にシンガポールに着く
木に行灯を吊るす
本番のパレード前、面出教授からのオリエンテーション

Musashino Art U. design studio students prepare materials
for the installation by making holes in plastic bottles
300 plastic bottles used to create blue cages of light,
in cooperation with I Lohas
On the grounds of Musashino Art U
Picking up materials after arrival in Singapore
Hanging lanterns from trees
Orientation from Prof. Mende before the parade

学生は東京で準備したインスタレーション用の器材を持ち込み、シンガポールの日本人幼稚園児のための行灯づくりワークショップを行い、当日のパレードや灯篭流しなどのパフォーマンスに協力した。開発著しいシンガポールの夜景に、「あかりを消してスローな夜を」というコンセプトが束の間ではあるが共鳴した瞬間だった。

Students prepared candle installations in Tokyo and flew all of the necessary equipment to Singapore with them. In Singapore, they held a lantern making workshop at a local Japanese Kindergarten and on the night of the event lead a Children's Parade and float lanterns in the bay. In the significantly growing and developing city of Singapore, the event concept of "Turn off the lights and take it slow," may be just a fleeting moment, but the idea resonates loud and clear for that instant.

クリフォート・ピアから湾に放たれた青い灯篭
Little blue cages of light released in the water from Clifford's Pier

道端のキャンドル・インスタレーションを楽しむ人たち
Visitors enjoying the candle installations

ワークショップで子供たちが作ってくれた手持ち行灯
Hand-held lanterns made at the children's workshop

子供たちを指導するゼミ生
Students are playing with kids

インスタレーションを興味深そうに覗き込む子供たち
Children peering at the installations with great interest

子供たちが描いた行灯のための絵
Lantern drawings by children

キャンドル・キオスクでは、通行人の人たちにも
即興で制作参加していただいた
At the candle kiosk anybody interested is welcome to make a lantern

光で世の中を感じる　佐藤卓

面出さんとの出会いは、1997年に開館した安曇野ちひろ美術館の仕事依頼の時でした。当時から面出さんはグラフィックデザイナーの福田繁雄さんらと共にこの美術館を運営する財団の理事をされていて、開館にあたり新しいロゴマークを制作したいと、財団の方と私の小さな事務所にお出でいただいたのです。それまでにも、もちろん面出さんのことはよく存じ上げていて、数々の仕事も拝見しておりましたが、突然直々に来ていただくということになって、改めて自分の事務所の照明がまったく恥ずかしいものであることに気づき、お出でいただく直前に「しまった！」と思ったのを今でもよく覚えています。

その後も、いろいろなところでお世話になっていますが、ここ数年ご一緒しているものに、毎年夏至の日に近い土曜日の夜に、原宿の表参道で行われる「キャンドルナイト」があります。これは、電気を消してキャンドルの光を楽しみながら、そもそも人の営みにとって光とは何か、そしてエネルギー問題など広く想いを馳せる、地域と共に社会へ広げていくプロジェクトです。面出さんは文化人類学者の竹村真一さんらと共に、最初にこのプロジェクトを立ち上げたメンバーでした。そこに私も途中から声を掛けていただき、ここ数年参加してきました。

電気をふんだんに使う世界の都市が、同じ日の同じ時間に電気を消していけば、地球の半分の夜の都市がキャンドルの光になりうつり変わるという壮大なスケールをイメージしてこの活動が始まったことは、想像に難くありません。光の害と書いて「光害」という言葉を初めて使ったのは面出さんでしたが、日本の都市は大量の蛍光灯が天井に貼りつくコンビニエンスストアーなどが象徴的であるように、下品なまでに明るすぎるのです。そのために物の陰影はなくなり、店頭では陰のない食品などが記号のように人に届く。そして唾液も出さずにものを食べる。もともと日本の住環境は、障子やふすまなどに代表されるように、淡い光が自然と物との関係をひとつにし、そこから人の感性が育まれました。それが戦後になって、合理主義に技術革新が拍車をかけ、暗いよりも明るい方が豊かであるかのような擦り込みがなされ、日本から暗い闇が消えたのです。これが、日本人がそもそもの日本を見失い始めた時だったのです。

面出さんは、光で世の中を感じている人です。このような日本の状況をどれほど憂いていることか、本人に聞かずとも心が痛むほど今になってわかるのです。「キャンドルナイト」は、文明とは何かを考えさせる、小さな光による大きなプロジェクトだと思っています。そこにお誘いをいただいたことに感謝いたすと共に、面出さんには光で世の中を見る術を教えていただきました。「キャンドルナイト」の夜に合わせ、オリジナルのキャンドルを毎年制作する美術大学やデザイン学校の学生達も、間違いなく私と同じことを面出さんから教わっているに違いありません。この、光というものがいかに人の思考や文明にとって大切であるかに気づくタイミングは、若ければ若い時ほどいい。面出さんに習った学生達が、これからの日本の光をもっともっと素敵にしてくれるだろうと、大いに期待しています。

佐藤 卓　*Taku Satoh*

グラフィックデザイナー／1955年東京都生まれ。1981年株式会社電通入社。退社後、1984年に佐藤卓デザイン事務所設立。商品デザイン、子ども教育番組のアートディレクション・総合指導などを手掛ける。武蔵野美術大学デザイン情報学科客員教授。東京ADC・東京TDC・JAGDA・日本デザインコミッティー・AGI会員。21_21 DESIGN SIGHT ディレクター。

Graphic designer/Born in Tokyo in 1955. Joined Dentsu Inc. in 1981. After quitting the company, founded Taku Satoh Design Office Inc. in 1984. Engaged in product design and art direction and total guidance of educational programs for children. Visiting Professor, Department of Design Informatics, Musashino Art University. Member of Tokyo ADC, Tokyo TDC, JAGDA, Japan Design Committee and AGI. Director, 21_21 DESIGN SIGHT.

Sensing the World through Light

I first met Mr. Mende when I was asked to do a job for the recently opened Chihiro Art Museum Azumino. He and the graphic designer Shigeo Fukuda were both directors of the foundation administering the museum, and Mr. Mende and a representative of the foundation arranged to come to my small office about a new logo they wanted to have made. I had of course known about Mr. Mende and seen a number of his works. I realized how disgraceful the lighting was in my office, now that he was about to visit the place in person, and well remember even now how panicked I felt just before he arrived.

He has helped me in many situations since then. In recent years, I always see him at an annual event called "Candlenight" held on Omotesando in Harajuku on a Saturday night close to the summer solstice. It is a project meant to encourage more people in society to think about a wide range of questions such as what light means to work and energy-related issues; it is also an opportunity to turn off electric lights and enjoy the light of candles. Mr. Mende and Shin'ichi Takemura, the cultural anthropologist, began this project. An invitation was extended to me, and I have participated for the last several years.

If the cities throughout the world that use so much electricity were to turn off lights at the same time on the same day, the cities that are in nighttime in half the globe would suddenly become lit by candles. That would indeed be an event on a grand scale. It is not difficult to imagine how this movement began. It was Mr. Mende who first used the word *kogai* written with the characters for 'light' and 'harm', to mean the harmful effects of electric light. Japanese cities are brightly lit to the point of vulgarity; convenience stores for example have enormous numbers of fluorescent lights on their ceilings. As a result, things are robbed of any nuance, and shadowless food items that have been rendered into signs are served to customers. Food is made so characterless it is no longer appetizing. In the traditional Japanese residential environment, faint light introduced for example through shoji screens established a connection between nature and the man-made environment and nurtured a distinctive sensibility. Since the end of World War II, technological innovations have intensified a tendency toward rationalism, and the notion that 'bright' is preferable to, and signifies greater affluence than, 'dark' has become accepted. As a result, darkness vanished from Japan. That is when the Japanese began to lose sight of what Japan was like originally.

Mr. Mende is someone who senses the world through light. I am now aware of how much he fears these circumstances in Japan. I think of "Candlenight" as an important project that uses a little light to make us consider the nature of civilization. I am grateful to Mr. Mende for inviting me to this event and for teaching me the art of seeing the world through light. The students from art universities and design schools who make original candles for use on "Candlenight" no doubt have learned the same lesson from Mr. Mende that I have. The younger one is when one realizes how valuable light is to people's thoughts and to civilization, the better. I have great hopes that students who have learned from Mr. Mende will make light in Japan more wonderful.

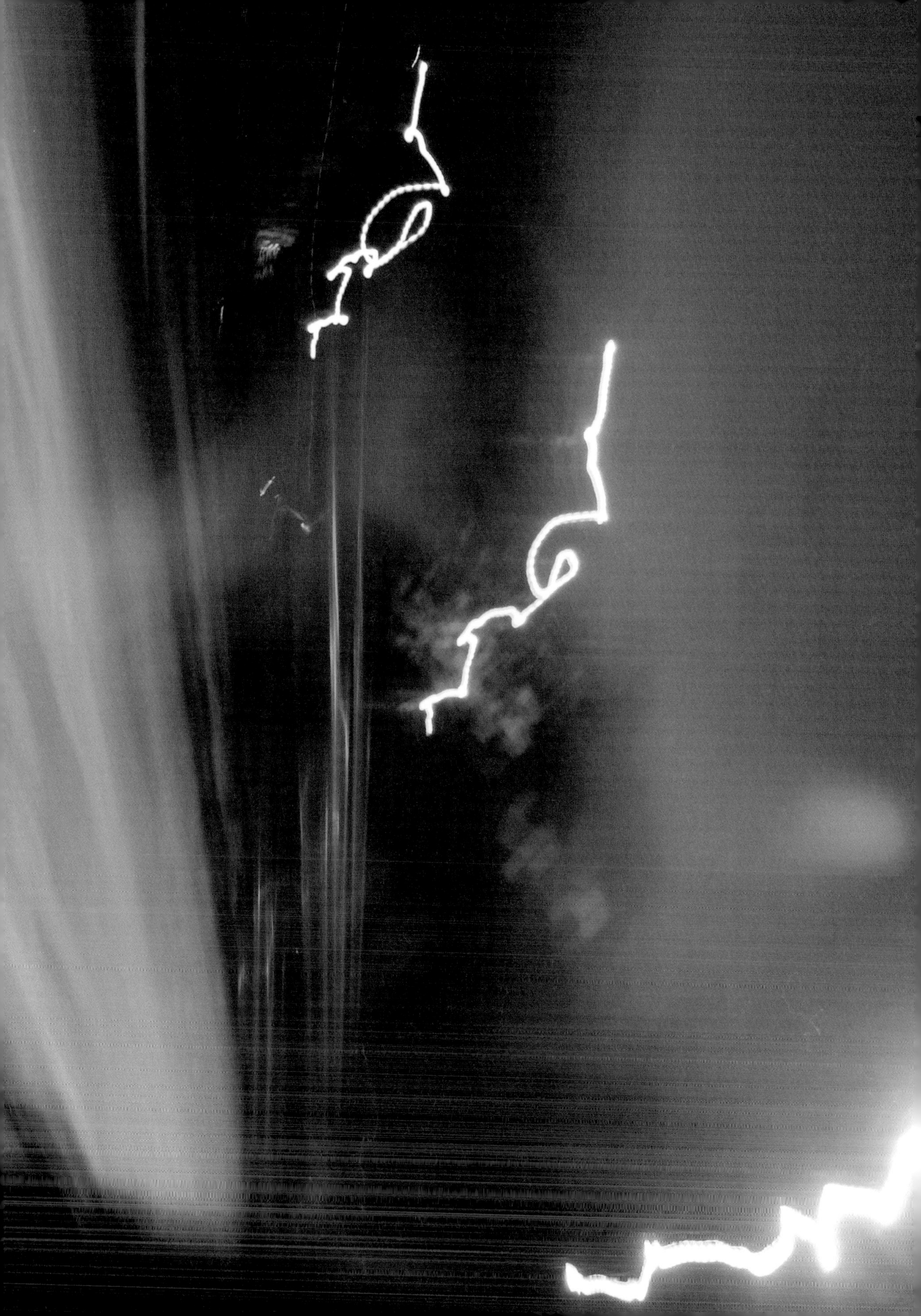

光を操る
MANIPULATING LIGHT

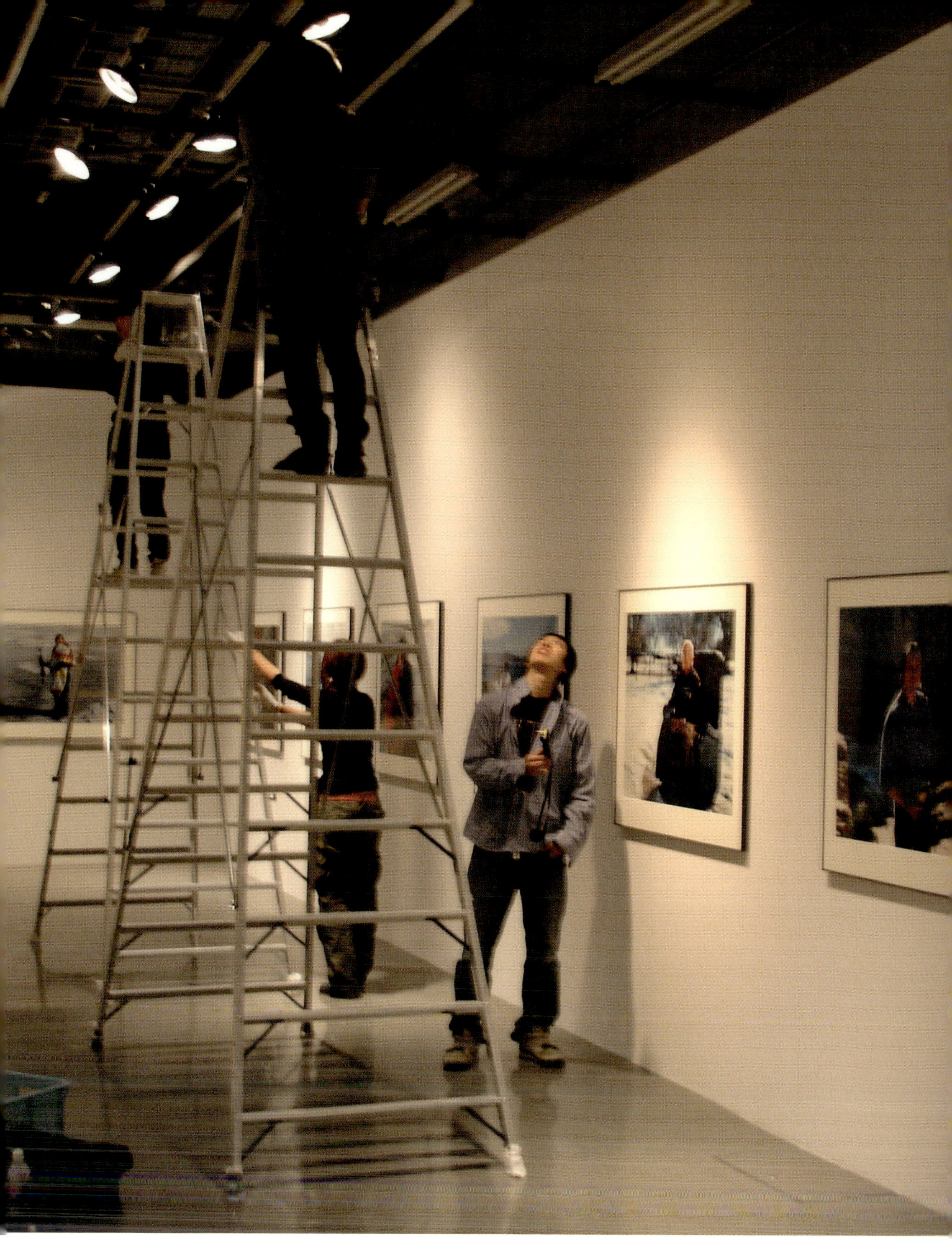

光を操る
MANIPULATING LIGHT

光のデザインを学ぶというプログラムの最後になってやっと、発想豊かにのびのびと内なるイマジネーションを全開する機会が与えられる。光を見ること、集めること、触れることを経験し、自然光、闇、陰影と対峙することで鍛えられ筋肉や感覚や知性が、光を操る段では自由な創作の糧になるに違いない。

光の創作課題はグループ課題と個人課題とに分かれる。ゼミの期間は3年後期から卒業まで18か月(1年半)なので、卒業制作に費やす6か月を抜くと残り12か月に最多で5課題ほどが出題されることになる。デザインの仕事は共同作業の中で成し得ることが多いので、私は常々チームの中のコミュニケーション能力を養うように指導する。

10年間のゼミ活動では、「佐賀町プロジェクト」「ライトアップ・ゲリラ」「キャンドルナイト@Omotesando」「明治神宮・闇のライトアップ」「岩室温泉郷照明計画」「新正卓展」「江戸東京たてもの園ライトアップ」「歌舞伎町ルネサンス」など、多種多様な光を操るグループ・プロジェクト課題をこなしてきた。また、個人課題としては「時をデザインせよ」「LEDを料理せよ」「癒しの空間をデザインせよ」「動く影をデザインせよ」「昼光をデザインせよ」「光と水をデザインせよ」「街かどの光を再生せよ」などを出題し、各自の創造性豊かな成果とプレゼンテーションを要求した。

学生の興味は都市環境、商業環境、住環境などに留まらず、ファッションと光を合成した「ウエアブルな光」、劇場パフォーマンスの要素の入った「光のインスタレーション」、光の芸術性をテーマにした「ライトアート」、そして映像や写真を駆使したものまで様々な展開を見せた。

最終的に個人が私の出題に対して回答する訳であるから、このような光を操るデザイン行為に対しては、私は厳しく批評する。学生が光のデザインを学ぶ過程での教育プログラムには責任を持つが、卒業制作を含む学生作品については自己責任を追求する。そこにはうまく行かないことも覚悟で自らを表現できるチャンスを与えるべきだと信じるからだ。光を操る(デザインする)という行為は、最終的には自分の創造力を過信しながら一気に仕上げるしか方法はない。ここでも許させる失敗をたくさん犯しながら成長していく訳である。

At the end of the program to study the design of light, students are given an opportunity to give full play to their own imagination. It is hoped that their muscles, sensibilities and intellects, disciplined through the experience of seeing, gathering and touching light and confrontation with natural light, darkness, shade and shadow, will enable them to engage in free, creative work now that they are called upon to manipulate light.

Creative work assignments are divided into group assignments and individual assignments. The seminar takes 18 months (that is, a year and a half), from the second semester of the third year to graduation. Since six months are spent on the graduation work, 12 months are left for, at the most, five assignments. Since most design work is the result of collaborative endeavor, I always try to nurture in students the ability to communicate within a team.

In ten year of seminar activity, we have undertaken diverse group projects in the manipulation of light such as the Sagacho Project, Light-Up Guerilla, Candlelight@Omotesando, Meiji Shrine—Light up the Darkness, Iwamuro Spa Lighting Project, Taku Aramasa Exhibition, Edo-Tokyo Open Air Architectural Museum Light Up, and Kabukicho Renaissance. Individual assignments have included "Design Time," "Deal with LEDs," "Design a Space That Comforts," "Design a Moving Shadow," "Design Daylight," "Design Light and Water," and "Revive the Street-corner Light" and demanded of students highly creative works and presentations.

Students have been interested in more than just the urban, commercial and residential environments. They have produced a wide range of projects including "wear-able lights" synthesizing fashion and light, "light installations" incorporating aspects of theater, "light art" having as its theme the artistic character of light and projects making use of film and photography.

I am severe in critiquing such designs that manipulate light since ultimately these are the responses of individuals to my assignments. I take responsibility for the educational program that offers students a process for studying the design of light, but they must take responsibility for their own work including their graduation work. I believe students ought to be given the opportunity to express themselves, but they must also be prepared for the possibility that things will not turn out well. There is ultimately no other way to manipulate (that is, design) light than to act with confidence and decisiveness. Permitting students to make many mistakes enables them to mature.

新潟岩室温泉 —— 光のまちづくり
Iwamuro hot spring, Niigata
Town planning by light

この計画は武蔵美が新潟県岩室温泉と共同して取り組んだ「アートとデザインによる街づくりプロジェクト」の一環として参加したものである。2006年よりは文部科学省現代GP(教育ニーズ取り組み支援)プロジェクトとして位置付けられ、計画が本格化していった。

　取り組みの分野はストリートファーニチャー計画、観光デザイン計画、サイン計画、キャラクター開発などに及び、学内では通称「いわむろみらい」創生プロジェクトと呼ばれた。私たちのゼミが担当したのは「光の環境デザイン計画」である。私たちは豊かな自然と美しい闇を背景にした岩室温泉郷のあちこちに、芸術的な光景を出現させようとした。都会に失われたさりげなく芸術的な情景が、ここ岩室にたくさん潜ませることを期待した。

We participated in the project as part of the "community planning with art and design" project, a campaign the Musashino Art University and Iwamuro spa resort in Niigata Prefecture jointly launched.

　The project covered street furniture planning, tourism design planning, sign planning, character development and other elements. The project was referred to as the "Iwamuro mirai" creation project in the university. Our seminar assumed responsibility for "light environment design planning". We tried to create artistic scenes everywhere in the Iwamura spa resort against the background of rich nature and beautiful darkness. We hoped that numerous naturally artistic scenes that were lost in urban areas would be hidden in Iwamuro.

岩室温泉郷の全域に夜景イメージ図を描いてみた
A nightscape image drawing of the entire Iwamuro Hot Springs area

上 | 昼の岩室の風景は何とも穏やかで
　　 のんびりしている
下 | メインストリートの現状の光環境は
　　 機能的ではあるが特徴にかけていた
T: *The relaxing daytime atmosphere of Iwamuro*
B: *The main street lighting environment is functional, but unique*

　プロジェクトは先ず現地の光環境調査から始まり、コンセプト・デザインの立案、照明手法の開発、現地での照明実験、光環境マップの制作、最終プレゼンテーション、という順序で進められた。
　プレゼンテーションでは「岩室あかり宣言」を採択し今後の開発を期待した。
　＊岩室あかり宣言
　岩室は暖かいあかりでもてなします。
　岩室は美しい闇と自然を大切にします。
　岩室にはまぶしいあかりがありません。
　岩室では光のアートに出会えます。
　岩室は美味しいあかりを目指します。

The project proceeded from the light environment surveys at the site to the development of concept design, development of lighting methods, on-site lighting experiments, preparation of light environment maps and final presentation. In the presentation,
　"Iwamuro Light Declaration" was adopted in hope of subsequent developments.
Iwamuro Light Declaration
　1　Iwamuro treats visitors with warm light.
　2　Iwamuro values beautiful darkness and nature.
　3　Iwamuro has no glaring light.
　4　You can meet art of light in Iwamuro.
　5　Iwamuro seeks tasty light.

光を操る
MANIPULATING LIGHT

1　岩室温泉で岩室の方たちに昼間プレゼンテーションを行う大学での発表と違って緊張感がみなぎる
2,3,4　地元に残る源泉「霊雁の湯」での照明演出。キャンドルも使って印象的な景色を創りたいと思った
5　メインストリートでの光環境調査から始まった

1　Presenting to the local people of Iwamuro in Iwamuro Hot Springs. This is different then a presentation on campus and the students are nervous
2,3,4　A candle illumination for the source of the hot spring, "Reigan Hot Spring"
5　We started with a lighting environment survey of the main street

上｜夜間には真っ暗で、まったく観光資源になっていなかった「下の堤」という名の池
中｜「下の堤」に魅力的な夜景を与えるための照明実験を行った
下｜「霊雁の湯」の照明計画イメージ図

T: At night it is too dark for the popular pond "Shimono Tsuzumi" to even be a tourist attraction
M: A lighting experiment to make the nightscape more attractive for "Shimono Tsuzumi"
B: An image rendering for "Reigan Hot Spring"

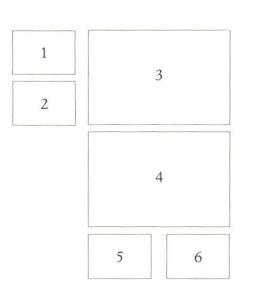

1 たくさんの岩室提灯を皆で制作する
2 岩室を訪れるお客様に提灯でのそぞろ歩きを
  楽しんでもらうことを提案した
3 実験的に夜道を提灯で歩いてみる
4 メインストリートに面した「高島屋」の
  照明計画イメージ図
5 計画案を打合せにはたくさんのスケッチが
  飛び交った
6 気分転換に、LEDを手にして夜の公園での光遊び

1 Everyone worked together to make Iwamuro lanterns
2 We proposed the lanterns for added enjoyment
  to leisurely strolls for tourists who visit Iwamuro
3 Experimentally, we walking down the dark streets
  holding lanterns
4 An image rendering for "Takashimaya"
  on the main street
5 Exchange of idea sketches at a team meeting
6 For a little diversion, light play with LEDs
  in a local park

原寸大の「いわむろ行灯」を持ち込んで
メインストリートで照明実験を行った
*A lighting experiment on main street
using a full-scale size Iwamuro Lantern*

光を操る
MANIPULATING LIGHT

新正卓展照明計画
Light planning for the exhibition by Taku Aramasa

武蔵美を退任する写真家・新正卓の展覧会が学内9号館の地下展示室を使って開催された。新正先生の希望で繊細でコンセプチュアルな展示照明計画を面出ゼミが担当した。照明デザインの仕事で最も難易度が高く、腕を試されるのが美術館照明であり展示照明である。私たちは作者の意図を十分に反映しながら更に快適で印象的な照明デザインを提案した。場の連続性（シークエンス）を考慮しながら劇的に、しかも繊細な写真芸術の妙を損なわない展示照明がテーマであった。

学生はプロさながらの責任感をもって、クライアントとの企画会議からコンセプトの立案、照明手法の開発から器具の手配、施工管理までを高密度な時間の中で体験する。絵に描いた餅ではない現実の成果が得られることに、これらの実践的プロジェクトの意味がある。

An exhibition was held in a basement gallery in building No. 9 of the university for Mr. Taku Aramasa, a photographer who was resigning from the faculty of Musashino Art University. At the request of Professor Aramasa, the Mende Seminar was in charge of the lighting planning for the exhibition. We fully respected the intention of the photographer and proposed a comfortable and creative light environment.

In lighting planning, the sequence in the field was taken into consideration. Exhibition lighting was sought that was dramatic but would never ruin the skills of delicate photographic art. Students experienced plan meeting with the client, concept development, development of lighting methods, acquisition of equipment, and construction control in highly dense timeframe with a sense of responsibility as if they were professionals. Obtaining actual results rather than the pie in the sky is the key to practical projects at real-world project sites.

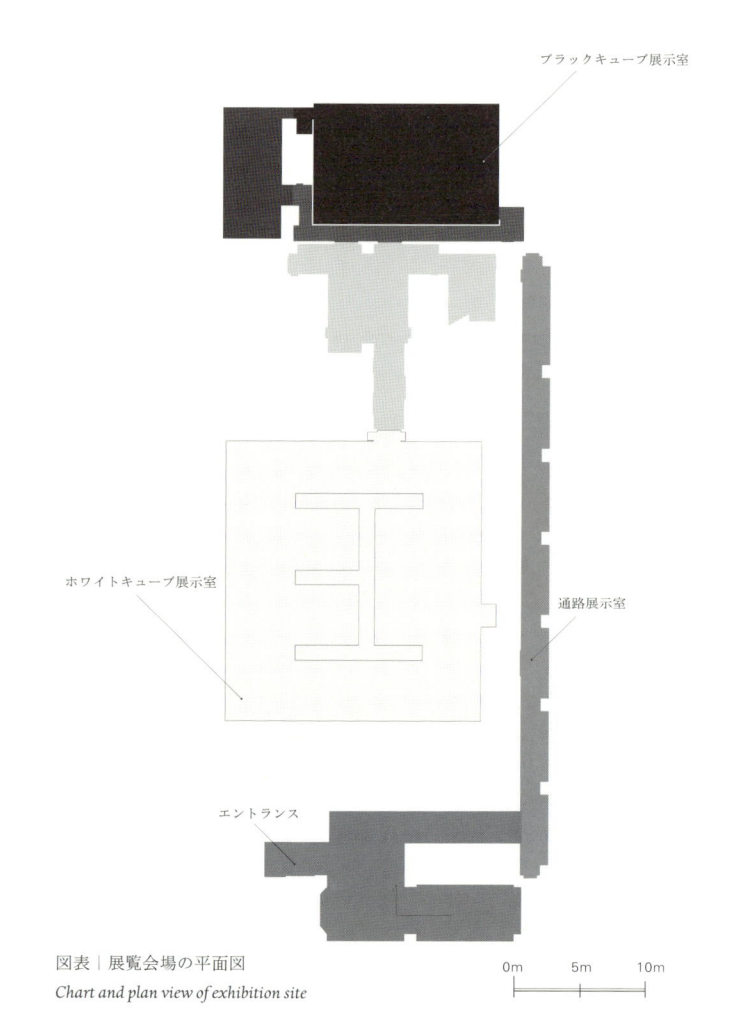

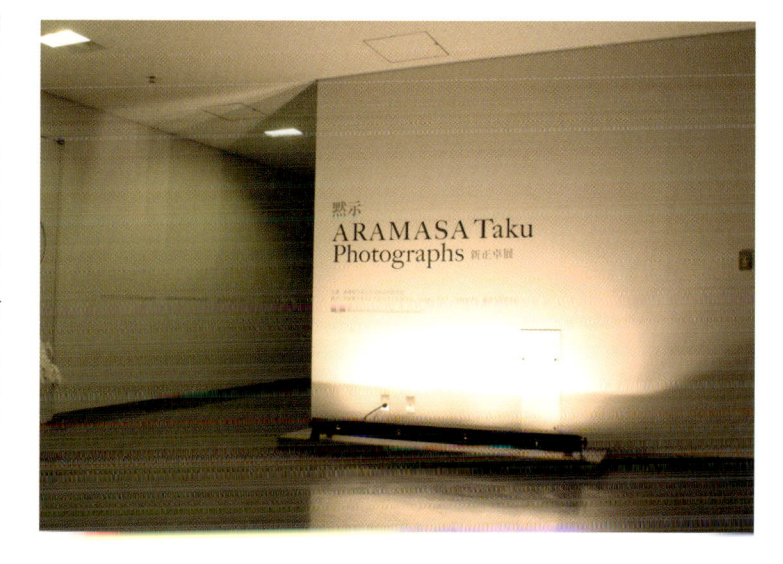

図表｜展覧会場の平面図
Chart and plan view of exhibition site

エントランス部のアップライト演出
Lighting at the entrance

ホワイトキューブ展示室は壁面を均一に照明し、
ニュートラルな光環境を計画した

Wall surface was uniformly lit in the white cube exhibition room, and neutral light environment was planned

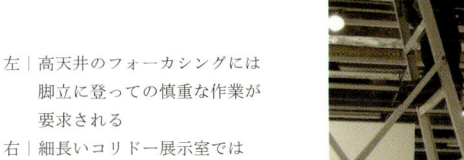

左｜高天井のフォーカシングには
　　脚立に登っての慎重な作業が
　　要求される
右｜細長いコリドー展示室では
　　リズミカルな演出照明にした

*L : Focusing light on the high ceiling
　　required careful work on a stepladder
R : Rhythmical lighting was applied in
　　the long and thin corridor exhibition room*

光を操る
MANIPULATING LIGHT

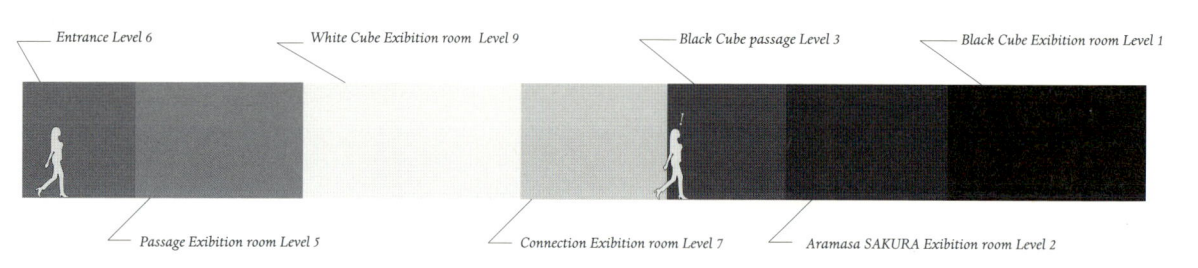

ブラックキューブ展示室にはレンズ付きのカットスポットと
青色カラーフィルターを装着した12㎜管蛍光ランプを採用した
Lens-equipped spotlights with reduced hot-wire and 12-mm fluorescent lamps
with a blue color filter were adopted in the black cube exhibition room

Entrance Level 6 — White Cube Exibition room Level 9 — Black Cube passage Level 3 — Black Cube Exibition room Level 1

Passage Exibition room Level 5 — Connection Exibition room Level 7 — Aramasa SAKURA Exibition room Level 2

展示会場は徐々に暗くなるようなシークエンスを計画した
The exhibition rooms were arranged in a sequence in which darkness gradually increased

SYMBOL	NAME	SPEC	ATTACHMENT
▶	Cut Spot light	日本応用光学 FPC-200w(Black)	東京舞台照明 NDフィルター #208 #209 #210
◯∪	12mm蛍光管	JEFCOM パラスリム	東京舞台照明 カラーフィルター #78

主要ディテールの検討会で黒板にスケッチをする
Sketches are made on the blackboard in the session for discussing major architectural details

間接照明用の部品をゼミ室で制作する
Parts for indirect lighting are made in a room of the seminar

足元の僅かな隙間に間接照明を施工する
Indirect lighting is applied in small gaps at the foot

さて本番の照明調整作業。ここで大切な作品を前に緊張感がはしる
Lighting is tuned for real exhibition.
Tension is created in the presence of actual valuable work

光を操る
MANIPULATING LIGHT

江戸東京たてもの園
ライトアップ

江戸東京たてもの園は江戸・東京の歴史的な建築を保存展示する目的で作られた屋外博物館である。宮崎駿（スタジオジブリ）の「千と千尋の神隠し」のモチーフとして園内の下町仲通や銭湯（子宝湯）が使われたことでも有名だ。

2010〜2012年まで面出ゼミの最後の3年間、たてもの園の依頼によって「紅葉とたてもののライトアップ」をゼミ課題として取り組んだ。私たちに求められたことは、樹木の豊富な園内の紅葉とともに、歴史的な建築群を夜間に演出し、多くの来園者から好評を得ることである。

何事も初年度には苦労が付きまとう。19名のゼミ生がいたので全員参加でことを乗り切ったが、園内全域のライトアップ作業は膨大だった。しかもぎらぎらと明る過ぎない照明をテーマとしキャンドルによる露地行灯なども多用したため、毎日の点滅監理にも時間がかかった。次年度は8名のゼミ生だったのでやり方を工夫して、下町仲通のたてものを主役にした5分間のライトアップ・パフォーマンスを行ったり、来園者が家族で楽しめるような影絵遊びなどを仕込ませた。最終年はゼミ主体ではなく、空間演出デザイン学科の1〜3年の有志参加を募り、30名余りの学生による人海戦術で乗り切った。

これほど広い敷地を相手に実践的な照明計画を行うことは滅多にない。11月末の寒さとともに、上手く行かなかったことの反省こそ貴重である。もちろん上手く行ったことで少しばかり褒められることが、全ての勲章になった。

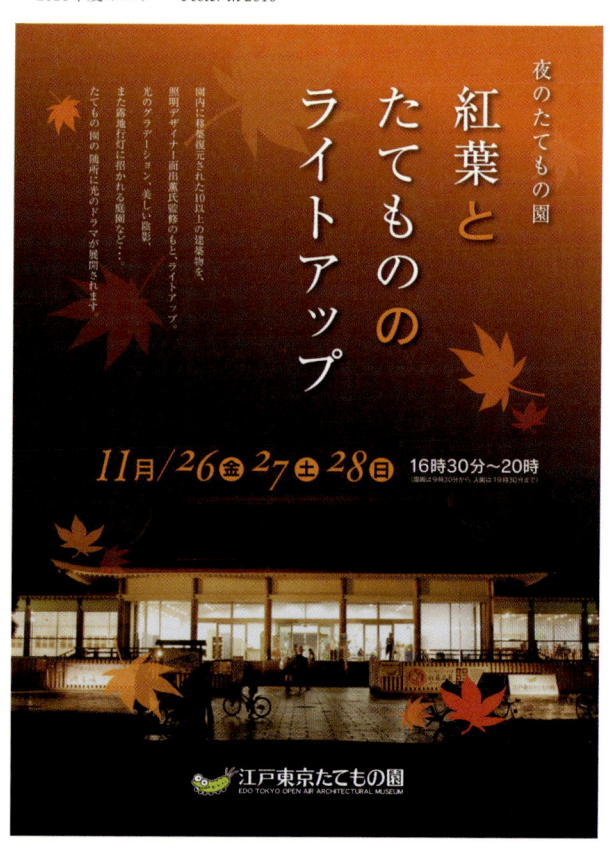

2010年度ポスター　Poster in 2010

Edo Tokyo Open Air Architectural Museum
Light up

The Edo-Tokyo Open-air Architectural Museum was constructed to preserve and exhibit historic buildings in Edo and Tokyo. It is well known that some structures that appeared in Spirited Away, an animation film produced by Hayao Miyazaki of Studio Ghibli, were designed based on shitamachi-nakadori, or street in downtown, and the public bathhouse (kodakara-yu) in the Open-air Museum.

The Mende Seminar students worked on the seminar assignment, "Light-Up of autumn leaves and buildings" for the last three years of the seminar in 2010 through 2012 as requested by the Museum. The requirements given to us were to dramatize during the night a group of historic buildings as well as abundant trees at the premises with reddening leaves for winning the appreciation of numerous visitors.

Every endeavor involves hardships in the initial year. We overcame the difficulty with the participation of all 19 seminar members. Light-up throughout the Museum required a huge amount of work. Lamps with a candle and paper shade were used frequently along small alleyways under the theme of illumination that was not excessively glittering. As a result, turning on and off the lights every day took much time. In the second year, eight seminar members were available. Five-minute light-up performance was conducted mainly for the buildings along shitamachi-nakadori. Playing with shadow pictures was designed to entertain family visitors. In the last year, a labor-intensive method was chosen with nearly 30 students. Our seminar did not play a central role but invited volunteer first- through third-year students in Kuude, or Department of Scenography, Display & Fashion Design, Musashino Art University.

Practical lighting plans in such a wide area are seldom implemented. The cold of late November and reflection on failures were of high value. A little admiration for successful jobs was our reward.

2011年度ポスター　Poster in 2011

2012年度ポスター　Poster in 2012

約7ヘクタールの敷地を持つ園内には、江戸時代から昭和初期までの20棟前後の復原建造物が建ち並び、11月末〜12月初めには紅葉の季節を迎えてたくさんの方々が夜のライトアップを楽しみに来園する。

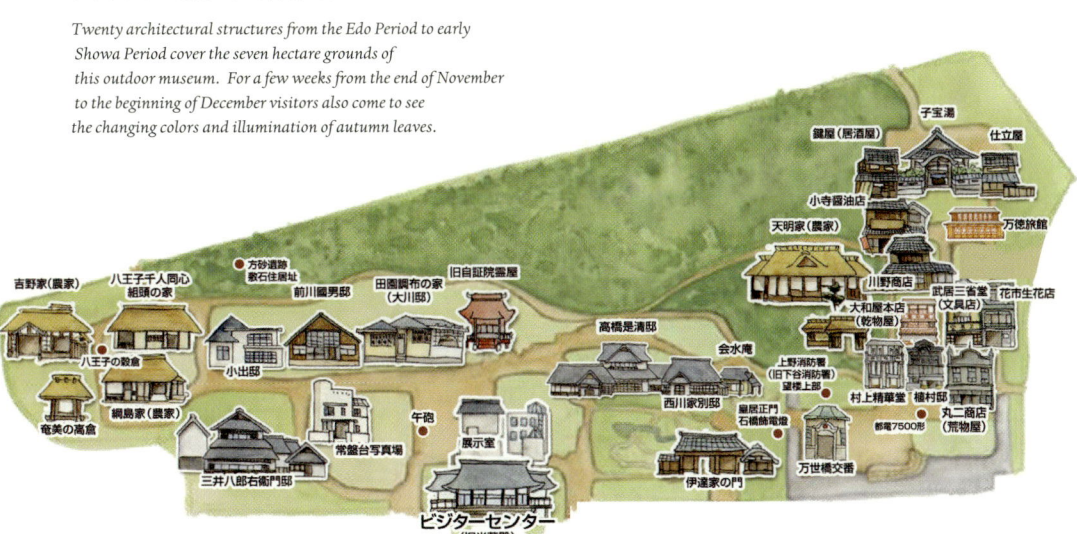

Twenty architectural structures from the Edo Period to early Showa Period cover the seven hectare grounds of this outdoor museum. For a few weeks from the end of November to the beginning of December visitors also come to see the changing colors and illumination of autumn leaves.

小金井カントリークラブ

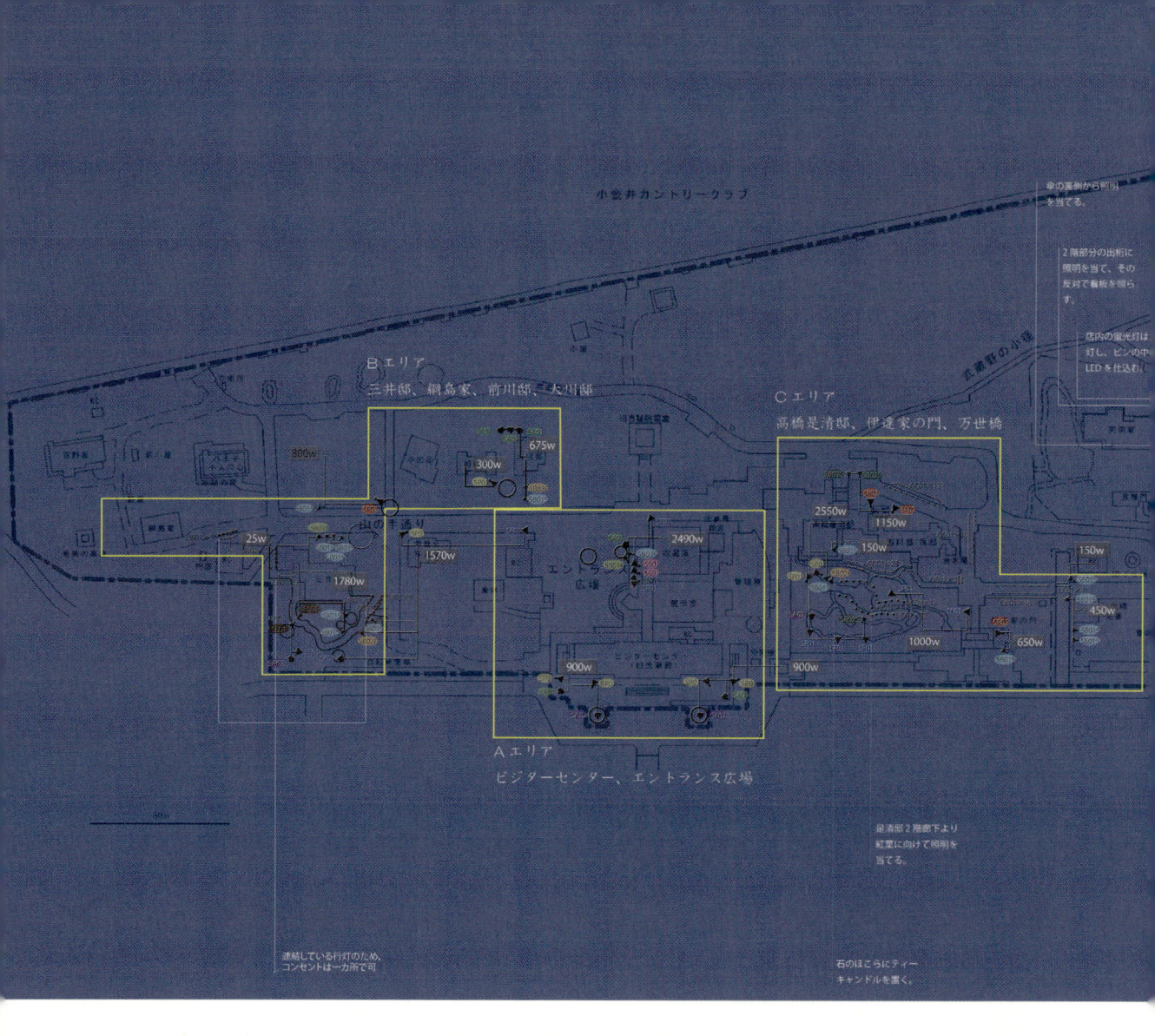

Bエリア
三井邸、綱島家、前川邸、大川邸

Cエリア
高橋是清邸、伊達家の門、万世橋

Aエリア
ビジターセンター、エントランス広場

傘の裏側から照明を当てる。

2階部分の出桁に照明を当て、その反対で看板を照らす。

店内の蛍光灯は灯し、ピンの中にLEDを仕込む。

最清邸2階部下より紅葉に向けて照明を当てる。

連結している行灯のため、コンセントは一カ所で可

石のほこらにティーキャンドルを置く。

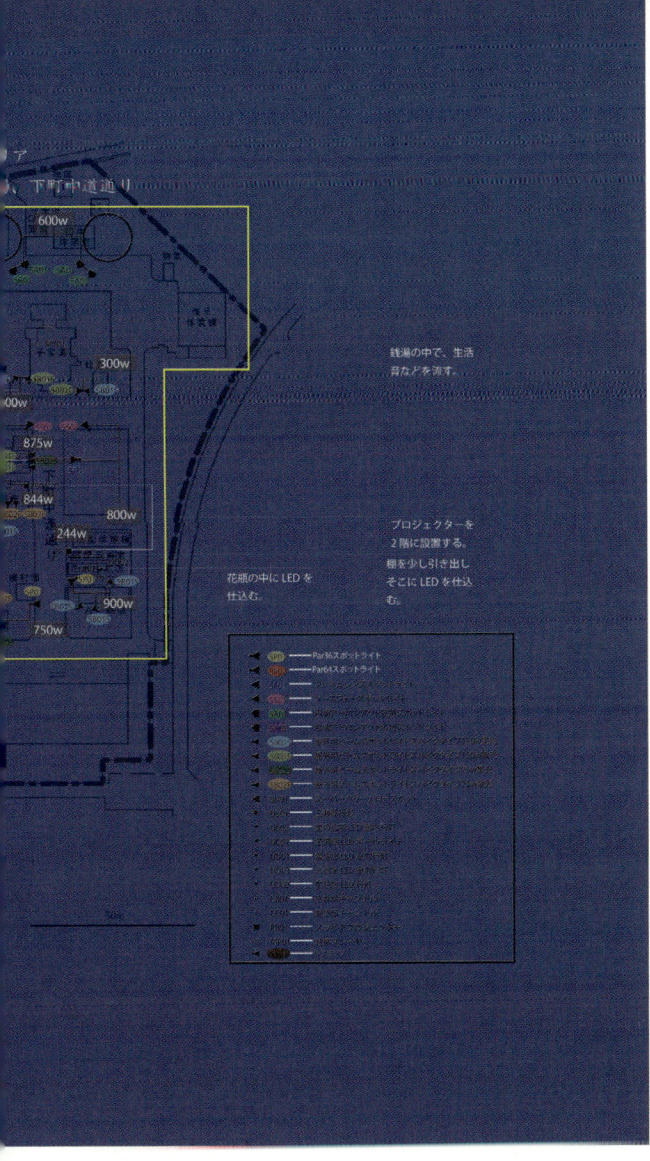

私たちは広大な敷地と施設を、1 エントランス＋ビジターセンターエリア、2 三井邸＋綱島家エリア、3 高橋是清邸＋伊達家の門エリア、4 下町仲通＋子宝湯エリア、の4エリアに分割して、それぞれにライトアップの考え方と照明手法を展開して行った。また園内の保安灯も、ライトアップの演出効果を損なう場所23か所に限り消灯をお願いした。しかし同時に足元の暗さを解消するために、柔らかい光の保安灯をデザインし設置して行った。

　環境に優しいライトアップは明る過ぎず暗過ぎず。照明計画が持つ治安や防犯の役割と、美しく心にしみる夜景創出を両立させる難しさも学習した。

We divided the wide premises and facilities into (i) entrance and visitors' center area, (ii) Yamate-dori street and the house of the Mitsui family, (iii) house of Korekiyo Takahashi and the gate to the house of the Date family and (iv) shitamachi-nakadori and east square, and developed the light-up concept and lighting method for each area. Numerous lamps were generally designed to be turned on in the premises. It was requested that lamps were tuned off at 23 locations where they were likely to spoil the effect of light-up. At the same time, handmade lamps with a paper shade were installed as security lamps with diffusion of light not hitting the eyes of people because excessive darkness at visitors' feet was not allowable.

　Environmentally friendly light-up should use neither too bright nor too dark lights. Students learned how difficult it was to have lighting plans play the role of ensuring security and preventing crimes, and create beautiful and touching nightscape at the same time.

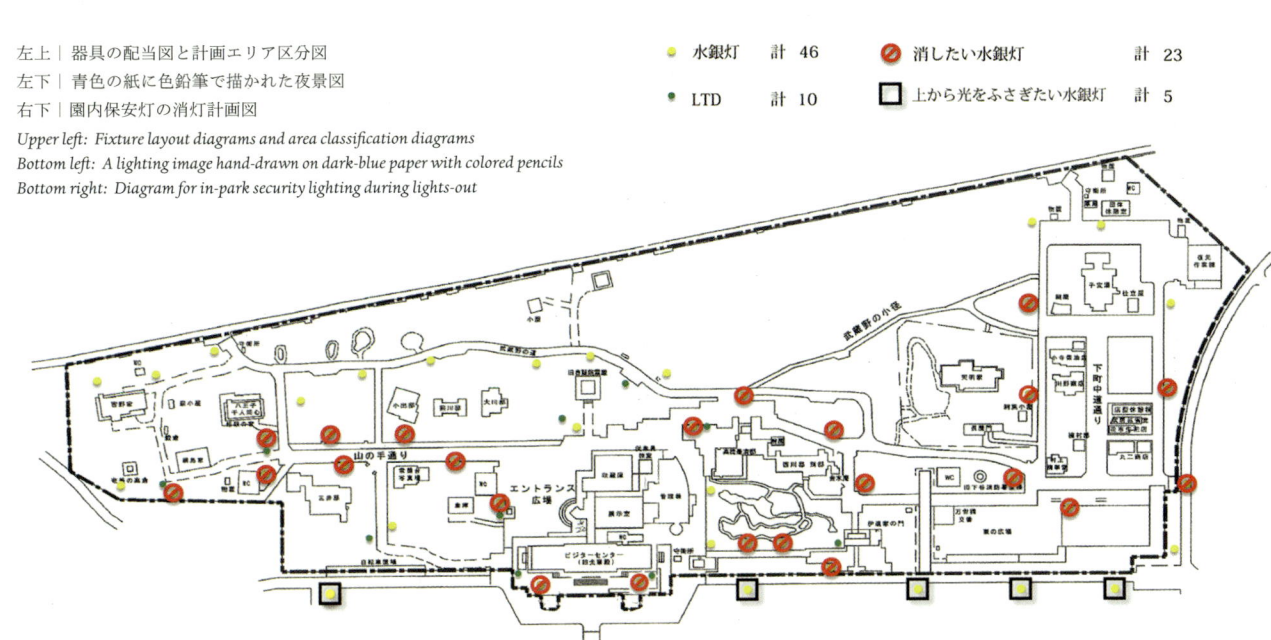

左上｜器具の配当図と計画エリア区分図
左下｜青色の紙に色鉛筆で描かれた夜景図
右下｜園内保安灯の消灯計画図

Upper left: Fixture layout diagrams and area classification diagrams
Bottom left: A lighting image hand-drawn on dark-blue paper with colored pencils
Bottom right: Diagram for in-park security lighting during lights-out

水銀灯　計 46
LTD　計 10
消したい水銀灯　計 23
上から光をふさぎたい水銀灯　計 5

MANIPULATING LIGHT

ビジターセンターの照明
Lighting of visitors' center

お客様を迎える第一印象は、エントランス外観と受付のあるビジターセンターだ。大屋根を持つ立派な建築は室内から溢れる光を大切に扱いながら、エントランスに追加されたお迎えのための行灯が計画された。センター内の天井照明はほとんど消灯し、逆に大きな天井をスクリーンとしてお迎えの映像を映し出した。

The first impression is given to visitors by the façade of the museum entrance and the visitors' center with the front desk. The lights bursting from inside the magnificent building with a large roof were carefully handled. At the same time, adding candlelight with a paper shade was planned at the entrance for welcoming visitors. The landscape created by the day-night reversal was symbolic. Most of the ceiling lights in the visitors' center were tuned off and welcoming images were projected on the ceiling used as a screen.

三井邸の庭園照明
Lighting of the garden of the house of the Mitsui family

三井邸には秋に燃えさかる見事な紅葉が点在している。中でも門柱付近で出迎える紅葉と、庭園内を回遊する時に楽しむことのできる紅葉は圧巻である。三井邸の室内から溢れる蛍光灯などの不要な光を先ず丁寧に制御した。更に回遊式庭園の流れに沿って露地行灯を設置しながらドラマティックな紅葉のライトアップを行った。

Spectacular blazing autumn foliage is scattered in the garden of the house of the Mitsui family. The leaves that greet visitors near the gateposts and those you can enjoy while going around the garden are the highlights. In the first place, unnecessary lights including those of fluorescent lamps pouring from inside were carefully controlled. Then, autumn leaves were dramatically lit while installing lamps with a candle and paper shade for small alleyways along the course of the stroll-through garden.

高橋是清邸の庭園照明
Lighting of the garden of the house of Korekiyo Takahashi

是清邸には紅葉だけでなく小川の流れや見事な苔を楽しむことができる。母屋の室内から庭を楽しむために配置された林の高木も、奥深い庭園のパースペクティブを作っている。これらを活かすために順路を占める露地行灯だけでなく、小川沿いの水を演出するLED照明や、高木を鋭くアップライトするためのスポット照明が施されている。

In the house of Korekiyo Takahashi, visitors can enjoy not only foliage but also the stream flow and beautiful moss. The tall trees arranged to help enjoy the view of the garden from a room of the main house offer a deep perspective of the garden. In order to make use of these components, not only the lamps with a paper shade for small alleyways were installed along the tour course but also LED lamps dramatizing the streamlet water and spotlights for lighting the tall trees were applied.

下町仲通＋子宝湯の照明
Shitamachi-nakadori and kodakara-yu

「千と千尋の神隠し」の舞台となった下町仲通には様々な趣の深い商店が建ち並んでいる。たてものの外観とお店の中のディスプレイが照明の対象だ。私たちは極力照明器具を目立たせないような位置を選定し、外観の小さな部分を丁寧に照明して行った。中でも子宝湯は象徴的な存在なので、背景の森も含めて絵画的に演出した。

Various shops are lined along shitamachi nakadori that appeared in Spirited Away and visitors can see the goods on display. The exterior of each shop is also tasteful and serves as an important object for lighting. We positioned lighting devices so as to make them as unnoticeable as possible and carefully lighted small parts of the exterior. We dramatized kodakara-yu as a picturesque object against the woods in the background because it was something like a symbol.

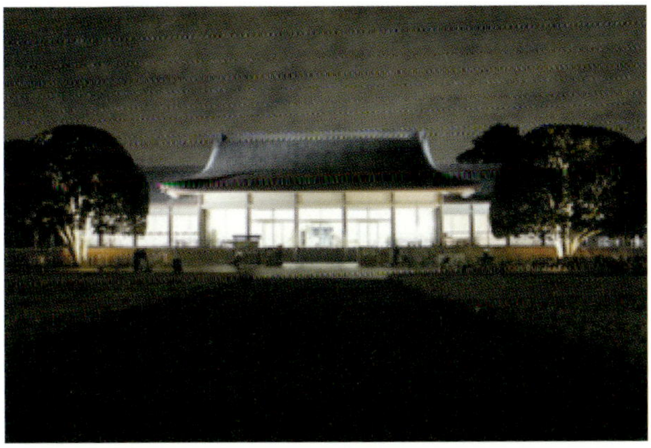

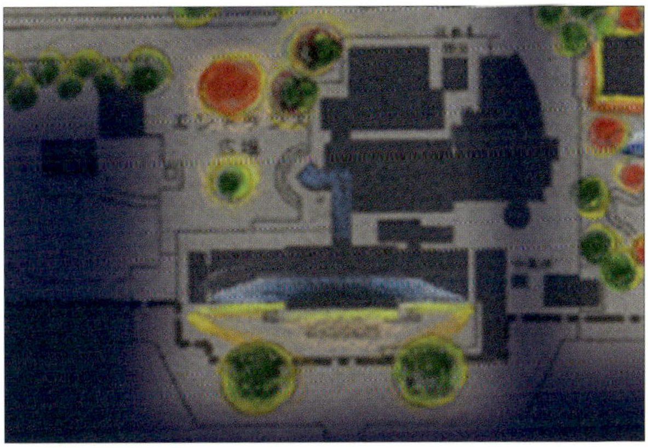

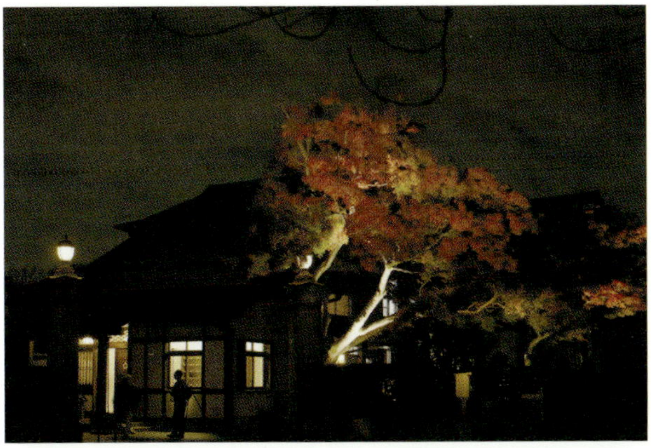

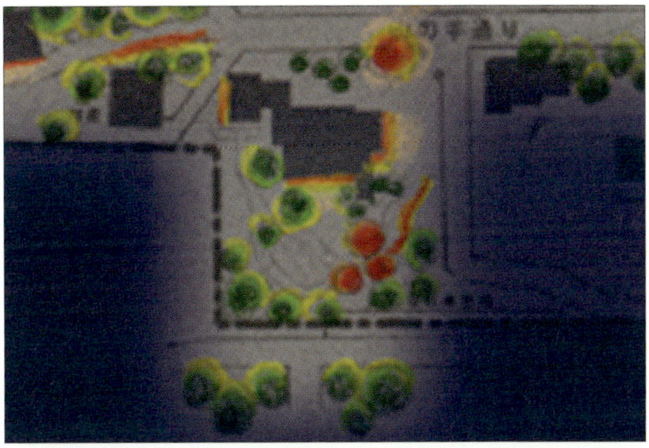

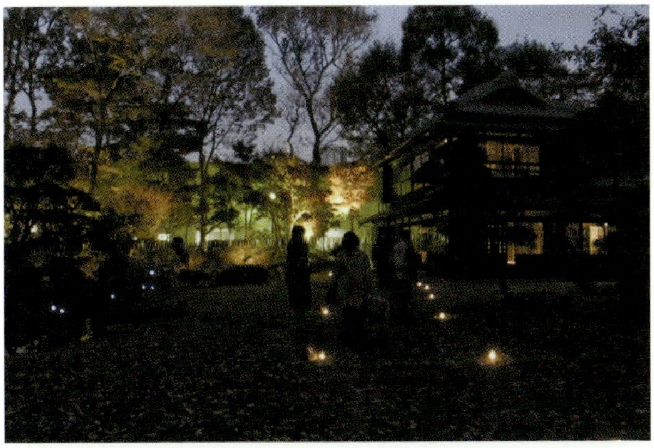

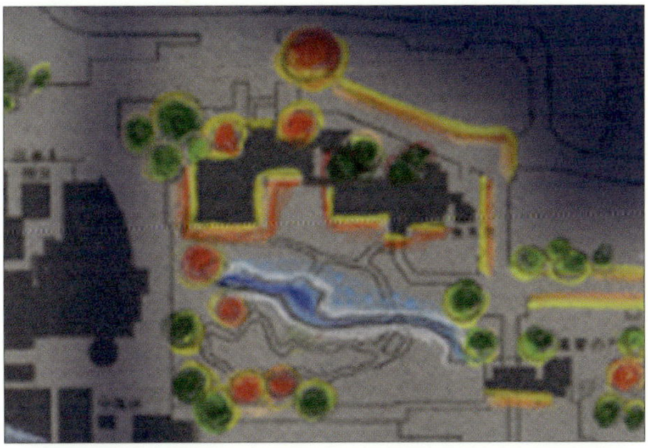

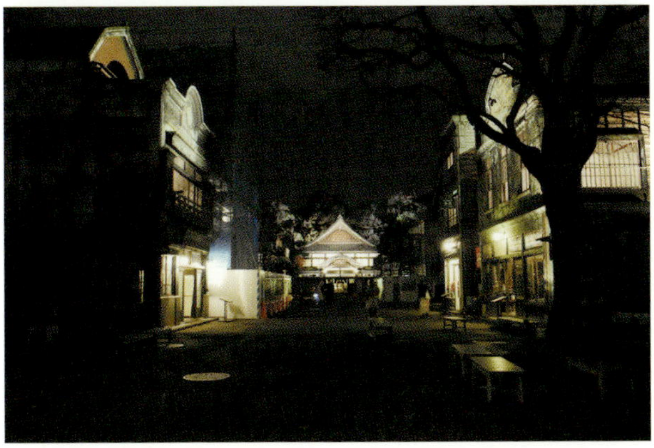

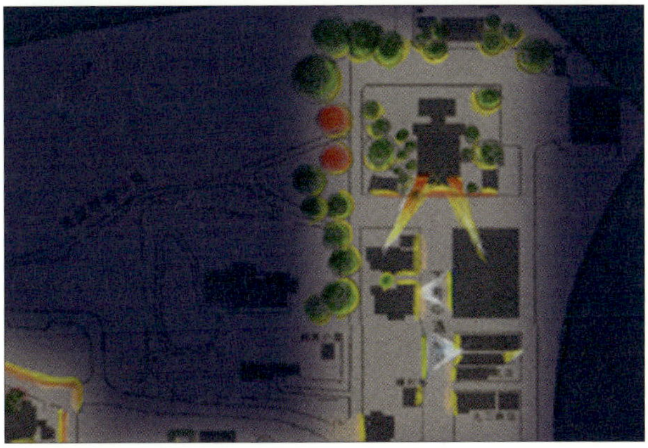

網島家の室内照明｜梁の通った古い木造民家では
ボランティアの方により囲炉裏に火がともされた
室内は天井へ柔らかい光を施した
*Tsunajima House Interior Lighting: A museum volunteer started a fire
in the open hearth of this old timber construction house.
Soft light illuminates the large beams and ceiling*

三井邸の紅葉照明｜鮮やかな紅葉が人々を招くように、
複数の角度からスポットライトを使用して、
陰影を意識した細やかなライトアップを行った
*Mitsui House Autumn Leaf Illumination: Several spotlights
from various angles were used to create shadowing
for this delicate illumination and the colorful leaves incite visitors closer*

是清邸の庭園照明
庭園では紅葉の他に小川の流れや
見事な苔を楽しむことができる
小川に水を感じるLED照明の
演出を行った
*Korekiyo House Garden Illumination:
In this garden a small river runs
into an autumn leaf pond
and beautiful moss carpets the banks.
LEDs are used to feel the presence
of water*

伊達家の門の照明
門には力強さを感じさせるために、
陰影の濃い景色を創りだした
*Date House Gate Illumination:
In order to fully experience
the dynamism of this gate,
we created strong contrasting palette
of light and shadow*

広場の電車
電車には異次元の
空間体験をしてもらうために、
LEDでほのかに青い空間を演出した
*Plaza Train Car:
In order to experience another dimension
of space in the train car
we added LEDs in a faint blue tone*

上｜村上精華堂は両サイドの
　　縦の外壁を強調した
中｜丸二商店＋花市生花店は
　　2階のオーナメントを照明した
下｜小寺醤油店は店内から漏れる
　　あかりを大切にした

T: We emphasized the façade on both sides
　　of the Murakami Seikado Store
M: The second-storey ornamentation
　　on the Marui Store and Hanaichi Seikaten
　　Florist is illuminated
B: Light glowing from the inside
　　of the Kodera Soy Sauce Shop is essential

下町通＋子宝湯｜下町仲通は金物屋や花屋など様々な商店が建ち並び、当時の店内風景を楽しめる。中でも子宝湯は象徴的存在で、背景の森も含めて絵画的に演出した。また、工事中のたてものには壁に移設前の姿やエピソード写真を投影した。

Old Town + Kodakara Public Bath: The main street of the old town has a florist, hardware store, and other merchants re-created for visitors to enjoy a retro atmosphere. Among all these buildings the most symbolic presence is the Kodakara Public Bath, as it stands picturesque in from of a forest of trees. Also, those buildings under construction have pictures of the building on the original site and other related snap shots hung for viewing.

下町仲通・光のパフォーマンス *Performance of light at shitamachi-nakadori*

2年目のたてもの園プロジェクトは主に下町仲通と東広場に集中して提案を行った。1年目のライトアップが大変な作業量に終始したので、2年目は「光の宴」というコンセプトを掲げて少し演出的に楽しめるライトアップ計画にしようとした。

その核になる企画が、子宝湯を中心とした下町仲通の全体を使った光のパフォーマンスだった。最終的には5分の音と光の屋外環境パフォーマンスを夜間30分ごとに開催した。下町仲通に位置する7戸の商店建築を使い、建築内部から発する光と、外観を照射する光の双方を細かく制御して、あたかもひとつひとつの建物がお互いに対話するかのようなストーリーを脚色した。観客には下町通を見渡せる道の中央に座り、両側から奥へと変化する光のパースペクティブを楽しんでいただいた。

劇場では舞台の上でおこる活劇を、100メートルもの街路全体を使って、しかも人海戦術の照明制御で行うのだから体力勝負の離れ業だったが、たくさんの観客を前にゼミ生たちは緊張と興奮を楽しんだに違いない。

We made proposals mainly on shitamachi-nakadori and in the east square in the second year of the Open-air Museum project. Light-up required a great amount of work throughout the first year. For the second year, light up plans were developed to help visitors enjoy dramatization under the concept of "feast of light".

At the core of the plans was the performance using light throughout shitamachi-nakadori with kodakara-yu playing the central role. Finally, five-minute environmental performance of light and sound was carried out at intervals of 30 minutes during the night. The lights from the inside of 7 shops located on shitamachi-nakadori and the lights that illuminated the exterior were finely controlled to develop a story in which the buildings had a dialogue with one another. Visitors, sitting at the center of the street commanding a view of the street, enjoyed a perspective of light that moved from both ends of the street into the background.

It was a difficult feat of physical strength because an action drama that would take place on stage in a theater was played on a 100-m-long street by means of labor intensive lighting. Students surely enjoyed the tension and excitement in the face of spectators

シーン1	日常的な光が消え、たてものたちに固有の光が宿っていく
シーン2	意思を持ったようにたてもの同士が反応して輝く
シーン3	ハンディキセノンによる素早い光の動き
シーン4	下町中通が非日常的で鮮やかな光により彩られる
シーン5	音楽に合わせたリズミカルな調光演出
シーン6	子宝湯がフルカラーLEDにより幻想的に変化していく

Scene 1	As daily light disappears, inherent light lives in each architecture
Scene 2	Buildings gleam, reacting to each other as if they intend to do so
Scene 3	Quick movement of light caused by handy xenon lights
Scene 4	Shitamachi-nakadori is highlighted by extraordinary brilliant light
Scene 5	Rhythmical light control to the music
Scene 6	Kodakara-yu changes fantastically using full-color LED

1	2
3	4
5	6

光を操る
MANIPULATING LIGHT

東広場での参加型照明演出　　*Participatory lighting in the east square*

2年目の東広場には、来園者が光や影と楽しく触れ合うことのできる装置が仕掛けられた。1つは土の広場に大きくしかれた光のカーペットだ。エプソン社から協力をいただいて7000ルーメンの強力デジタルプロジェクターを使用して、パソコンから様々に変化するグラフィック映像を投影した。時にゆったりと時に激しく変化する光の中に身を投じて、自らの影帽子と遊ぶ親子が印象的だった。

もう1つは藤棚に吊るされた白い布に投影されるカラーシャドー。RGB（赤緑青）3色のフィルターを使ったスポットライトの演出で6色の色のついた影を楽しむことができる。

In the second year, we installed devices in the east square to enable visitors to communicate with light and shadow. One was a light carpet laid on the soil square. A 7000-lumen strong digital projector that was presented through the cooperation of Epson Corporation was used to project ever changing graphic images from personal computers. It was impressive to see families playing with their own silhouettes in slowly and sometimes drastically changing light.

Another was colored shadows projected on white cloth hung under a wisteria trellis. Spotlights using red, green and blue filters generated shadows in six different colors.

181

左 | ビジターセンターの天井プロジェクションの事前実験
下 | 保安照明用の優しい行灯の試作実験と施工された山手通
L: A pre-event check on the visitor center roof projection
B: A prototype lantern for security lighting and construction along Yamate Street

全エリア合計器具リスト

器具番号	器具名称	型番	メーカーまたは調達先	ランプ	電気容量w	灯数	合計電気容量w	色温度
SP01	Par36スポットライト		川本舞台照明	シールドビームランプ300w	300w	11	3300w	2900K
SP02	Par64スポットライト		川本舞台照明	シールドビームランプ500w	500w	2	1000w	3200K
SF01	フレネルレンズスポットライト		川本舞台照明	ハロゲンランプ500w	500w	9	4500w	3050K
SS01	ソースフォーススポットライト		川本舞台照明	ハロゲンランプ500w	500w	3	1500w	3200K
SA01	岩崎アーバンアクト狭角スポットライト		川本舞台照明	メタルハライドランプ150w	171w	9	1539w	4500K
SA02	岩崎アーバンアクト広角スポットライト		川本舞台照明	メタルハライドランプ150w	171w	7	1197w	4500K
SB01S	屋外用ビームスポットライトスパイクタイプ	T4029B	ヤマギワ	シールドビームランプ150w集光	150w	22	3300w	2850K
SB01F	屋外用ビームスポットライトスパイクタイプ	T4029B	ヤマギワ	シールドビームランプ150w散光	150w	4	600w	2850K
SB02S	屋外用ビームスポットライトスパイクタイプ	T4029B	ヤマギワ	シールドビームランプ75w集光	75w	6	450w	2850K
SB02F	屋外用ビームスポットライトスパイクタイプ	T4029B	ヤマギワ	シールドビームランプ75w散光	75w	6	450w	2850K
SN01	スーパーベリーナロースポット		遠藤照明	CDM－T35w	40w	3	120w	3000K
UB01	三井邸行灯		武蔵美製作		5w	18	90w	
UB02	東の広場LED行灯		武蔵美製作		0w	666	0w	
UC01	是清邸LEDポールライト		武蔵美製作		0w	20	0w	
UD01	醤油屋LED室内行灯		武蔵美製作		0w	20	0w	
UD02	三省堂LED室内行灯		武蔵美製作		0w	20	0w	
UD03	生花店LED室内行灯		武蔵美製作		0w	20	0w	
CB01	三井邸キャンドル		武蔵美製作		0w	5	0w	
CC01	是清邸キャンドル		武蔵美製作		0w	41	0w	
PR01	スライドプロジェクター				244w	2	488w	
PC01	PC				60	2	120w	
MP01	音楽プレイヤー				0w		0w	
				エリア計			18654w	

Aエリア器具リスト

ビジターセンター

器具番号	器具名称	型番	メーカーまたは調達先	ランプ	電気容量w	灯数	合計電気容量w	色温度	
SP01	Par36スポットライト		川本舞台照明	シールドビームランプ300w	300w	4	1200w	2900K	スタンド
SF01	フレネルレンズスポットライト		川本舞台照明	ハロゲンランプ500w	500w	2	1000w	3050K	スタンド
SA01	岩崎アーバンアクト狭角スポットライト		川本舞台照明	メタルハライドランプ150w	171w	2	342w	4500K	ベース
SA02	岩崎アーバンアクト広角スポットライト		川本舞台照明	メタルハライドランプ150w	171w	2	342w	4500K	ベース
SN01	スーパーベリーナロースポット		遠藤照明	CDM－T35w	40w	2	80w	3000K	
				エリア計			2964w		

エントランス広場

SS01	ソースフォーススポットライト		川本舞台照明	ハロゲンランプ500w	500w	3	1500w	3200K	スタンド
SA01	岩崎アーバンアクト狭角スポットライト		川本舞台照明	メタルハライドランプ150w	171w	1	171w	4500K	ベース
SB01S	屋外用ビームスポットライトスパイクタイプ	T4029B	ヤマギワ	シールドビームランプ150w集光	150w	1	150w	2850K	
SB01F	屋外用ビームスポットライトスパイクタイプ	T4029B	ヤマギワ	シールドビームランプ150w散光	150w	1	150w	2850K	
SN01	スーパーベリーナロースポット		遠藤照明	CDM－T35w	40w	1	40w	4500K	
				エリア計			2011w		

Bエリア器具リスト

三井八郎右衛門邸

SF01	フレネルレンズスポットライト		川本舞台照明	ハロゲンランプ500w	500w	2	1000w	3050K	ベース：
SB01S	屋外用ビームスポットライトスパイクタイプ	T4029B	ヤマギワ	シールドビームランプ150w集光	150w	6	900w	2850K	
SB01F	屋外用ビームスポットライトスパイクタイプ	T4029B	ヤマギワ	シールドビームランプ150w散光	150w	1	150w	2850K	
SB02F	屋外用ビームスポットライトスパイクタイプ	T4029B	ヤマギワ	シールドビームランプ75w散光	75w	2	150w	2850K	
UB01	三井邸行灯		武蔵美製作		5w	18	90w		
CB01	三井邸キャンドル		武蔵美製作		0w	5	0w		
				エリア計			2290w		

前川邸・大川邸

SA01	岩崎アーバンアクト狭角スポットライト		川本舞台照明	メタルハライドランプ150w	171w	2	342w	4500K	スタンド
SA02	岩崎アーバンアクト広角スポットライト		川本舞台照明	メタルハライドランプ150w	171w	1	171w	4500K	スタンド
				エリア計			513w		

前川國男邸

SP01	Par36スポットライト		川本舞台照明	シールドビームランプ300w	300w	1	300w	2900K	ベース
				エリア計			300w		

大川邸

SB02F	屋外用ビームスポットライトスパイクタイプ	T4029B	ヤマギワ	シールドビームランプ75w散光	75w	1	75w	2850K

プロジェクトで使用した照明器具の仕様一覧表（部分）
A list of all the light fixtures used in this project

上｜下町仲通で隠れた位置から対岸の
  建物外観を照射する
下｜下町仲通のそれぞれの建物の室内照明を
  シナリオに合わせて変化させる

T: *We illuminated building facades on the opposite side
  of the old town main street from discrete places*
B: *To add an element of change to the old town,
  all the interior lighting is coordinated to a specific scenario*

歌舞伎町ルネッサンス
Kabukicho renaissance

歌舞伎町は「かぶく人々」の集う街である。かぶくとは「傾く」を語源として、自由奔放に振る舞うことであり精神の解放を起源とする。時に異様な身なりをしたり、人の目につく衣装を身に着けて目立つことを意味することもあるが、これが歌舞伎の起源だと考えると、歌舞伎町は芸術の街だともいえる。

芸術は自由奔放な人々の心に育まれ、歌舞伎町の芸術はとりわけ夜に開花する。これまでの猥雑な夜景を再起動しよう。シネシティ広場に可能な限りの静寂を取り戻し、ここに「かぶく人々」の精神を再現する。このプロジェクトは「光・再起動・かぶく人々」と題した光と影のパフォーマンス実験である。

Kabukicho is a meeting place for "Kabuku Hitobito".

Kabuku originates from the word "to incline" and means free-wheeling behavior. It evolved from the freeing of mind. The word sometimes means standing out clad in showy dress. If the word is the source of Kabuki, Kabuki-cho is considered a town of art. Art is nurtured by free-wheeling mind. The art of Kabukicho blossoms during the night in particular.

Obscene nightscape was re-initiated. Silence was regained inasmuch as possible in the Cinecity square and the mind of "Kabuku Hitobito" was reproduced. The project was an experimental performance of light and shadow entitled "Light, re-initiation and Kabuku Hitobito".

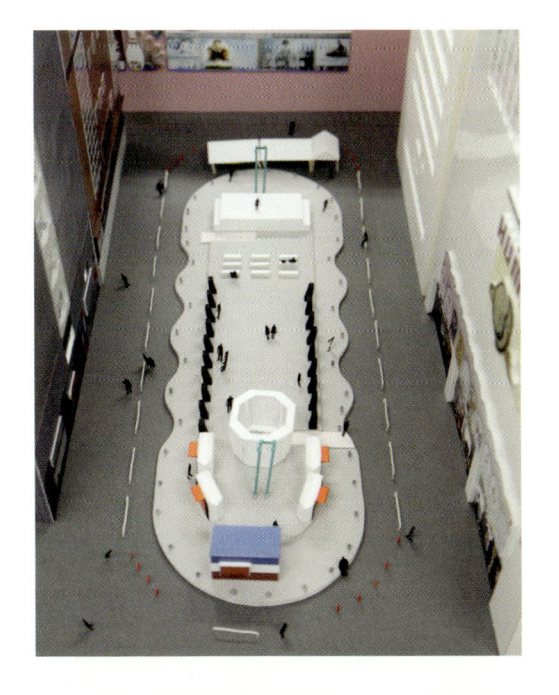

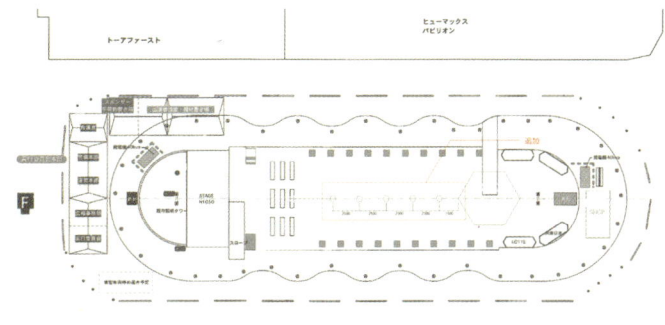

歌舞伎町ルネッサンス
ライティング・プロジェクト
光と影・かぶく人々
―

実施期間 | 2011年10月15日（土）、16日（日）の両日
　　　　　午後6:30〜8:00
実施会場 | シネシティ広場（コマ劇場わきの空地）
参加主体 | 歌舞伎町ライティング・プロジェクト実行委員会
　　　　　武蔵野美術大学 空間演出デザイン学科面出薫ゼミ
演出内容 | 15分のライティング・パフォーマンス
　　　　　「光と影・かぶく人々」
　　　　　1 影の塔（シリンダー幕の内部にうごめく陰影のアート）
　　　　　2 Kabuku-Hitobito
　　　　　（乱舞する人影／周囲のビル外壁に投影する陰影）
　　　　　3 光のオベリスク（雲を突く16000Wの光跡）
演出費用 | 100〜150万円
　　　　　（演出機材費、特殊照明レンタル費、運搬交通実費など）
主催 | 学生クリエータズ・フェスタ in 新宿 2011 実行委員会、
　　　新宿区
後援 | 経済産業省、文化庁、東京都、
　　　および各関連団体＋メディア各社

Lighting project/Light and shadow, and Kabuku Hitobito

Time: Saturday, October 15 and Sunday,
October 16, 2011 at p.m. 6:30 through 8:00
Place: Cinecity Square (empty lot near the Shinjuku Koma Theater)
Participants:
Kabukicho Lighting Project committee
Department of Scenography, Display & Fashion Design,
Musashino Art University
Mende Semina, Musashino Art University
Details: "Light and shadow and Kabuku-Hitobito"
(i) Tower of shadows (art of shading moving
in the membrane composed of cylinders)
(ii) Kabuku-Hitobito (shadows of dancing people
shading projected on the exterior walls of buildings in the vicinity)
(iii) Obelisk of light (towering trace of 16000-watt light)
Cost: One to 1.5 million yen (equipment, special lighting rental fee,
transportation, etc.)
Organized by: Executive committee of
Shinjuku College Creators Fiesta 2011 and Shinjuku Ward government
Sponsored by: Ministry of Economy, Trade and Industry; Agency
for Cultural Affairs; Tokyo Metropolitan Government;
organizations concerned and media organizations

周囲のビル大壁面に投影される「かぶく人々」のうごめく影
*On the huge sidewall of one of the buildings we projected "Tilting People,"
a collection of wiggling shadows*

シリンダー幕の内側でも、不可思議な影を創る
工夫がされている
*The inside of a cylinder curtain was rigged
to create mystical shadows*

周囲のビル壁面に映し出す人影のパフォーマンス（踊る学生）
Silhouettes of people dance on the side of the building wall (Dancing Students)

シリンダー幕の中には16KWのキセノンライト他の器材がぎっしり詰まっていて、様々な陰影や色の着いた光を演出した
Inside the cylinder curtain is a 16kw Xenon light along with a few other fixtures to create different shadows and add color lighting to the performance

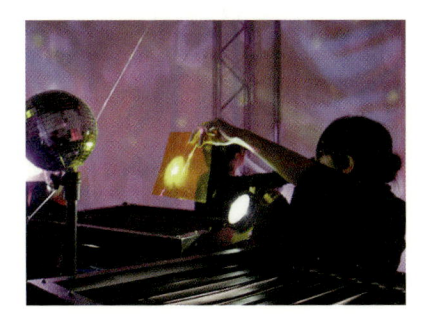

ミラーボールに向けてカラーフィルターを描けたスポットライトを調整する
Final checks on a spotlight fitted with a color filter shining on a mirror ball

光を操る
MANIPULATING LIGHT

完璧を作る隙間

小竹信節

滝野川という東京の下町で育った私は、それほど高くない電信柱からの、もうデコボコになったアルミ傘の付いた裸電球が作る長い影でよく遊んでいた。そんな実家の周辺の暗がりに似て、何て薄暗くて、幾つもの妖しい影の隙間があって"ときめく"んだろうかと思って、仕事で出向くたびに見つめていた東京国際フォーラムの建築照明デザイナーが彼だと知ったのは、大学に着任してからだった。

「この人、完璧だから!」と紹介された時、いちばん完璧などという幻想を口にしなさそうな、当時主任教授だった杉本貴志さんがいった意味が、やがてわかるようになった。過剰な明るさで人の感情を煽るのではなく、引き算の灯りに人の感覚は鋭敏にときめくのだ。そして、私と面出さんが育った場所は、さほど離れていないこともわかった。彼は、自ら完璧を作り出すのではなく、人がその灯りに身を置く時、それぞれの記憶を辿り出し加算されて行く中で、完璧に向かうそれぞれの隙間を埋め合わすのだ。

私の専門は舞台美術で、短縮され、時には引き延ばされた空想の時間や風景を追うごとに、刻々と変化する灯りと長い間付き合って来た。照明の力によって、再び振り出しに戻るように、舞台美術の見え方が左右されることが多いのだが、舞台照明家には二つのタイプがいて、一つは物が存在しないと灯りを作れない人と、もう一つは空気そのものに灯りを放てる人がいる。劇的な灯りは、コンサートがそうであるように、色の変化や光に伴う影の強弱によって煽ることで、これは真っすぐな道を進むように、それ程難しい作業ではないが、物が作り出す具体性とは別に、視点の合わない気配を作り出せる灯りには、にじり寄って来る霧で囲まれたように感情が揺れ、つまり、持って行かれるという想いと共に、何処とも知れぬ手掛かりのなかった記憶を呼び覚ますのである。

私が入って間もなく大学で作った授業の一つに、数分の短い物語として作り出す「影絵=影の劇場」がある。入って間もない新入生たちの、大学での最初の授業として、すでに15年程行って来たが、これは人間も物も影としてなら平等な姿に混じり合い、本来の姿とは別のマスキングされた存在として、人と既成物のみで輪郭を作り出し、"見立てる"ことの想像力を養うために行っている。アンデルセンが描いた『影法師』のように、影が自らの自由を得るために、邪魔になった本来の影の持ち主である主人を消し去るような姿に似ている。

当初、光と影の明快な輪郭を作ることによって、より的確な表現としての影の演出が主体だったが、それぞれが異なる分野での授業間の行き来が大変なことは承知で、私はある時、彼をこの授業に誘った。そこから学生らの表現にとても"不明瞭な"影の世界が妖しく顔を出し始め、江戸の影絵を引用して来たにも関わらず、ある意味で西欧の影絵の、いわば灰色を消し去った姿から、『陰影礼賛』のごとく、彼らの意識の中で徐々に、光と影の境界が蝋燭の灯りのように豊かに揺らぎ出したのである。面出さんが大学を去ってしまうことへの私たちの揺らぎと共に、見届けて来た陰影は、立ち会うことのできた幸運な学生全ての記憶の中に、しっかりと刻まれていますよ。

小竹信節　Nobutaka Kotake

舞台美術家、アートディレクター/1950年東京都生まれ。1975年から1983年まで演劇実験室「天井桟敷」の美術監督として舞台美術及び映画美術を担当。その後、数多くの舞台美術を手掛け、1991年度青山ワコール・スパイラル・ホールの芸術監督に就任し、装置のみによる演劇を試みる。造形作家として自動機械をテーマとした海外企画展に多数招待出品。武蔵野美術大学空間演出デザイン学科主任教授。

Stage art designer and art director/Born in Tokyo in 1950. Engaged in stage and film art design in 1975 through 1983 as an art director of "Tenjo-sajiki", or gallery, experimental theater. Subsequently involved in stage art design on numerous occasions. Assumed a post of art director of Wacoal Spiral Hall in Aoyama in 1991 and tried dramas exclusively using stage installations. Presented numerous works as a formative artist at overseas events as a guest under the theme of automatic machinery. Head of the Department of Scenography, Display & Fashion Design, Musashino Art University.

The Gap That Creates Perfection

Raised in a shitamachi district of Tokyo called Takinogawa, I used to play in the long shadows created by light from a naked light bulb under a beat-up aluminum cover set atop a not so tall utility pole. With its low-lit environment and mysterious, shadowy gaps that remind me of the darkness around my childhood home, Tokyo International Forum has fascinated me every time I visit it in connection with my work. It was only when I took up my position at the university that I found out that Mr. Mende had been the architectural lighting designer for the forum.

In introducing Mr. Mende, Mr. Takashi Sugimoto, who was then chairman of the department, said, "This man is perfect." Mr. Sugimoto seemed the person least likely to talk about illusions such as perfection, but eventually I came to understand what he meant by the remark. Instead of stirring up emotion by means of excessive brightness, Mr. Mende works on the principle of subtraction; that is, he creates a feeling of anticipation by keeping light low. (I also found out that Mr. Mende and I grew up not so far from each other.) He himself does not create perfection. Instead, when people place themselves in a light he has created, memories are evoked and added to the mix, filling in the gap between reality and perfection in each person.

My specialty is stage art and I have long been accustomed to light changing from moment to moment as imaginary time or landscape is compressed or stretched. The appearance of stage art is greatly affected by lighting; the power of lighting is such that it can completely alter the effect of stage art. There are two types of stage lighting designers. Those of the first type can only create light if an object exists, and those of the second type cast light on air itself. Dramatic lighting for events such as a concert is created through changes in color or the intensity of shadow produced by light; it is not that difficult to achieve. However, light that is unfocused creates a feeling that is like being enveloped by creeping fog; it stirs the emotions, creating a sense of being carried away yet also awakening memories from who knows where.

One of the classes I started at the university soon after I began teaching is "Silhouettes—A Play of Shadows" which calls on students to create a short story several minutes in length. I have been teaching this class, the first that new students take, for 15 years now. In shadow, humans and objects are blended together as equals; they become different from the masked things they normally are. Outlines are created from people and ready-made objects, and the object is to develop in students the imaginative ability to liken one thing to another. Their task is not unlike that of "The Shadow" by Hans Christian Andersen who does away with his master to gain his own freedom.

The original aim was to create silhouettes with distinct outlines and thus render more accurate portrayals, but one time I invited Mr. Mende to this class, even though I knew it would be difficult for him to teach successive classes in different fields. From that time, a mysterious world of quite ambiguous shadows began to appear in the students' work. Though referencing silhouettes of the Edo period, it took on some of the qualities of Western silhouettes. As in Tanizaki's "In Praise of Shadows," the boundary between light and shadow began to blur and flicker like a candle in the consciousness of students. We are made anxious by Mr. Mende's departure from the university, but memories of those flickering shadows will long be engraved in the minds of students fortunate enough to have encountered them.

個人課題 —— 卒業制作
Assignments for individuals —— graduation work

個人課題としての出題は10年間を通じて首尾一貫したものではなかった。学生の顔色をうかがいながら、その時期、季節がらを反映して2〜4週間ていどで完成させるものが多かったが、中には「光の練習曲」というような1週間課題を3週連続で出題したこともある。

　思い起こすと学生の自由な発想を啓発したかったので、抽象的な課題が目立っていた。以下、そのいくつかの例である。

「時をデザインせよ」
「LEDを料理せよ」
「癒しの空間をデザインせよ」
「動く影をデザインせよ」
「昼光をデザインせよ」
「光と水をデザインせよ」
「街かどの光を再生せよ」
「音を光で視覚化せよ」
「日常をリ・デザインせよ」
「光の自邸をデザインせよ」

　私は各自の創造性豊かな成果とプレゼンテーションを要求した。どのように自作を伝えて表現するかに厳しかった。

　しかし、個人作品としての集大成は卒業制作である。4年後期の4か月半ほどがこれに当てられる。学生としては私から与えられた課題ではなく、自分で自分に課題を出すことを初めて強いられるのが卒業制作だ。大学生活4年間の集大成として取り組むわけだが、思いの深さを100パーセントの表現に結び付けられる学生はほとんどいない。しかしそれも自らの作品なのである。学生は卒業制作をポートフォリオの最終ページに掲げながら社会へと出陣していくのだ。

Assignments for individuals were not consistent throughout the ten-year period. Two- to four-week assignments were generally given according to the time of the year or season while judging the feelings of students. I once gave one-week assignments for three weeks in a row such as the "etude of light". Looking back, I notice abstract assignments. This is because I wanted to encourage students to go through free-wheeling thought process. Some are listed below.

Design time
Process LED
Design healing space
Design moving shadows
Design daylight
Design light and water
Regenerate light in street corners
Visualize sound using light
Re-design daily life
Design your house of light

I requested creative results and presentation from each student. I was strict with students with respect to presentation and expression of their work.

　Graduation work is the summarization of individual work. Students work on graduation work for nearly four and a half months in the second semester of their fourth year. Students are forced to give an assignment to themselves by themselves that I do not give to them for the first time in their university life. Students carry out graduation work as culmination of four-year life in university. Few students can fully express their deep feelings. Yet it is the work of their own. They go into the world with their graduation work on the last page of portfolio.

卒業制作 2005 **石見学—半実在**
Graduation Work 2005: Manabu Iwami
The Harf Reality

作者の石見は愚直に美しく真っ白な床、壁、天井を淡々と作り続けていた。思い込んだら迷わない姿勢が見事だと思った。しかしこの作品はどう親切に見立ててみてもジェームズ・タレルのイミテーションの域を脱していない。制作中に私は石見を「ジェームズ」と呼んで揶揄したりもしたが、彼は動じた風もなく自分の仕事に打ち込んだ。

私は時々「真似ること」の大切さを学生に諭している。手習いや学習というのは元来、何かを確実に真似ることから始まる。オリジナリティがあるやなしやを学生に問うつもりはないが、つまらないものを真似るのは下品な行為だ。

この作品は歩を進めて奥に入るにつれて自然光に近づき戦ぐ風に触れるようにできている。ジェームズの愛弟子が育っていることを嬉しくさえ思えてくる。

素材｜木材, 布, 昼光

Iwami, who created the work, was coolly and honestly constructing beautiful white floors, walls and ceiling. The unshaken attitude was remarkable. The work was, however, something like an imitation of James Turrell's work at best. I made fun of him during the construction by calling him James. He yet devoted himself to the work unmoved.

I sometimes talk to students about the importance of "imitating". Calligraphy and learning basically start with the imitation of models. I never ask students whether their work is original or not. Imitating poor models is, however, a indecent act.

The work was designed to place you closer to natural light and make you touch breathing wind as you step forward. I was even delighted that a favorite disciple of James was growing.

Materials: Wood, fabric and daylight

卒業制作 2012　刈谷康時

Asobi

Graduation Work 2012: *Yasutoki Kariya*

作者コメント｜インスタレーションを制作する上でのモットーは『精度に遊びはなく、あなたの感想に、心に遊びを』と掲げており、解説するのは避けたいと思います。なので、強いていうのならば、『アソビ・デザイン』となります。いろいろに感じて頂けると幸いです。

面出評｜この作品は11個の白熱電球が直線状に吊られていて左端と右端の1個が触れて残りの10個に衝突する、という極めて単純な仕掛けである。しかし思わず作品の前に佇んでしまう。その魔力を作っている仕掛けの1つが乾いた衝突音である。カチッ、カチッ、とメトロノームのような機械音を刻んでいる。そしてもう1つは衝突のエネルギーを視覚化するための閃光だ。この瞬き（Blink）のお蔭で11個の透明な電球が実は同一人物であるようにも見える。

インスタレーション／電球, 電線, solenoid
arduino programming／5000×2000×500

Comment of the student : The theme of creating the installation is "Bring asobi, or pleasure, to your impression and mind but no asobi, or allowance, in accuracy". The work is therefore a result of asobi design. I would appreciate it if you would get varying feelings.

Comments of Kaoru Mende : The work is composed of eleven incandescent lamp bulbs suspended from a line. The bulb at right or left end simply touches the remaining ten. I, however, could not help stopping in front of it. One of the keys to the magic was the dry sound of collision. Mechanical clicking sound was made as if by it were produced by a metronome. Another was the spark of light that visualized the energy of collision. The blink made the eleven transparent bulbs look like a single person.

Installation / electric bulbs, electric cables, solenoid,
Arduino programming / 5000 x 2000 x500

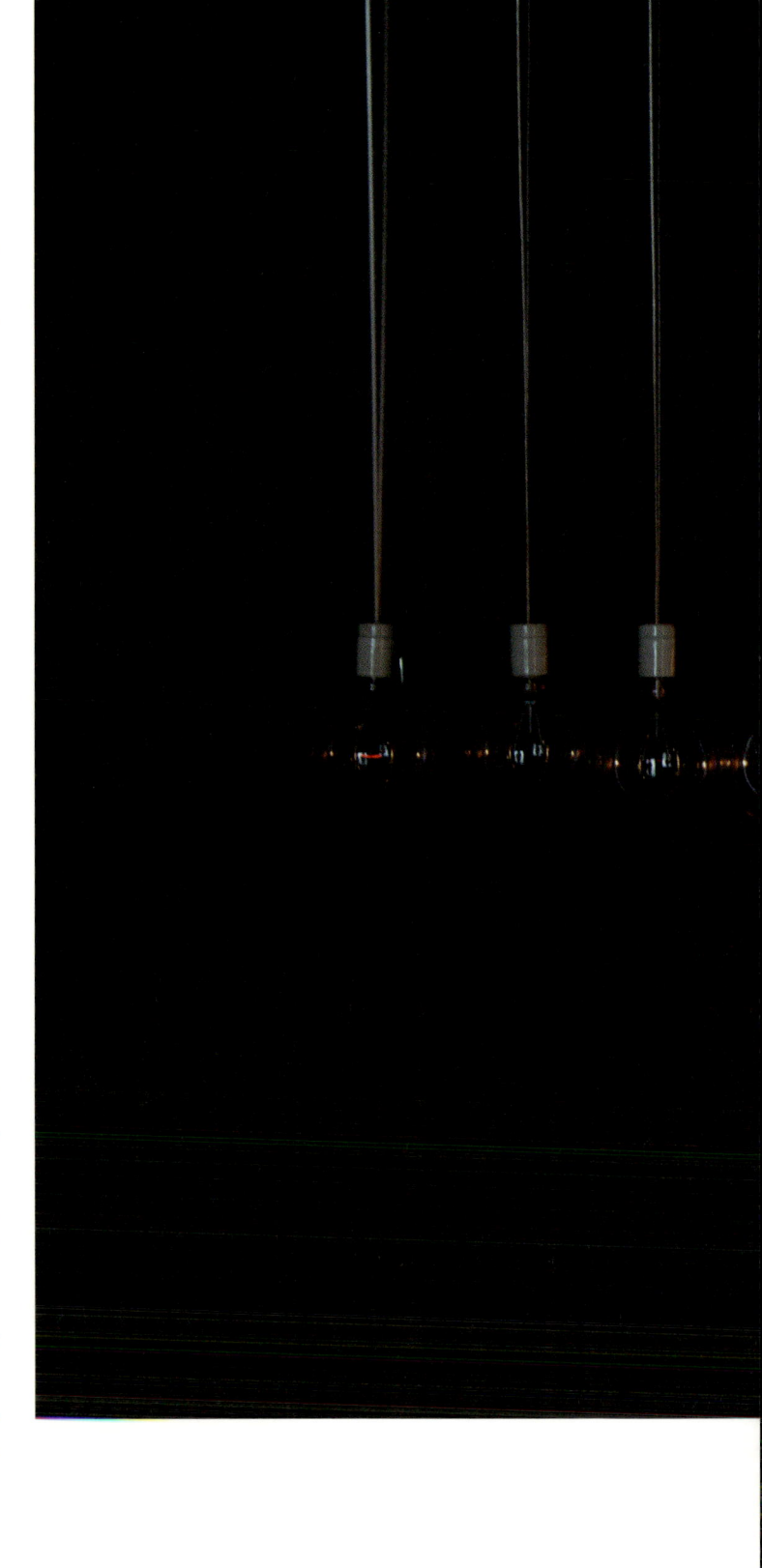

光を操る
MANIPULATING LIGHT

卒業制作 2007　長田直子
ひとつの線にある反対の場所

作者コメント｜これは陰翳を表現したものです。光のうつろいに目をやると、そこにあるたくさんのシーンを見ることができます。

面出評｜あらゆる光は常に変化の過程に存在する。一瞬の稲妻、飛び交う蛍、水面に揺れる夕陽などは早い変化なため解り易いが、忍び寄る夕暮れ、移り変わる季節感などにこそ変化の妙がある。今、光のデザインにはそのような時を伝える手法が必要だ。長田の作品が提供する4分間の光の変化は3分でも5分でもなしえなかった。空間を分割する2枚の霞網にクリプトンランプの光が明滅する。実に素朴な仕掛けだ。それはあたかも、私たちに天体の一部に暮らしていることを思い起こせといわんばかりに。

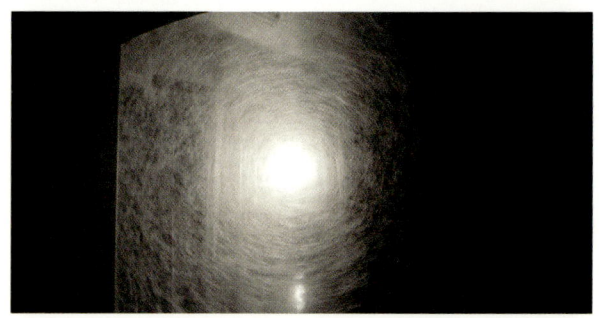

光の変化のしくみ｜半球形の明かりの周りを半球形のシェードが回転し、光の変化をつくりだします。1回転するのにおよそ4分かかります。

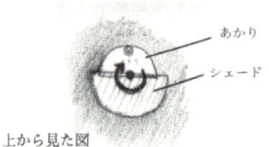

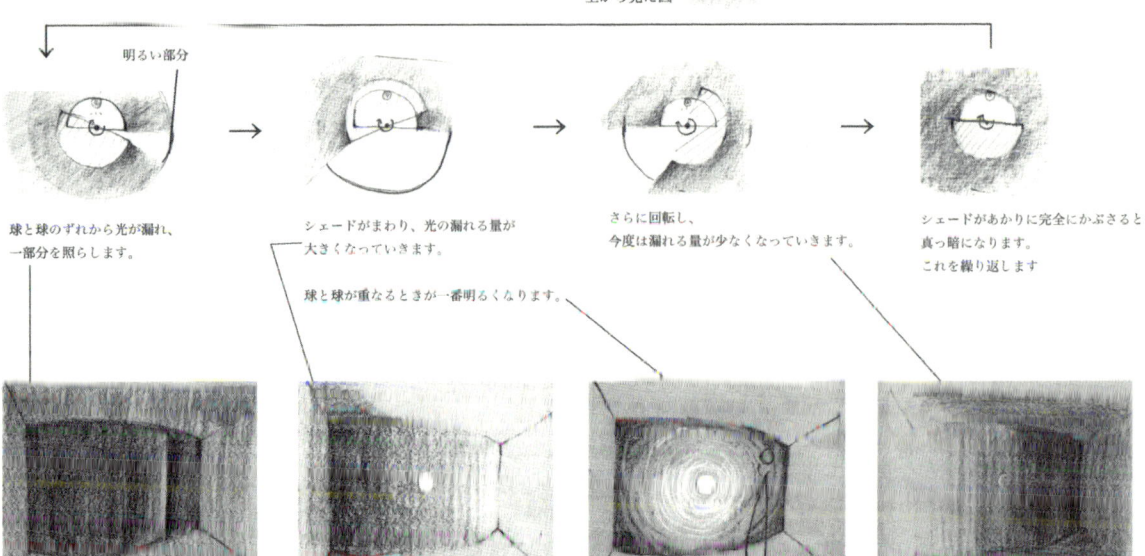

あかり
シェード
上から見た図

明るい部分

球と球のずれから光が漏れ、一部分を照らします。

シェードがまわり、光の漏れる量が大きくなっていきます。

球と球が重なるときが一番明るくなります。

さらに回転し、今度は漏れる量が少なくなっていきます。

シェードがあかりに完全にかぶさると真っ暗になります。
これを繰り返します

Graduation Work 2007: **Naoko Osada**
Place at the other end of a line

Comment of the student : This is an expression of shading. If you take a look at the changes of light, you may be able to see numerous scenes that exist therein.

Comments of Kaoru Mende : All types of light exist in the process of change. Passing lightning, flying fireflies and the sunset moving on the water are difficult to catch because of the speed of change. Sneaking dusk and the sense of changing seasons are examples of mysteries of changes. At present, methods of catching such points in time are required in the design of light. The change of light in four minutes that was presented by Osada may not have been achieved neither in three nor five minutes. Krypton lamps blinked in two pieces of fine netting used for catching small birds that divided the space. The extremely simple system was designed as if to remind us that we live in part of a celestial body.

Exhibition of a model, large space, hut, non-woven fabric, lamp and mixed media

卒業制作 2008　川島英明
灯り星

Graduation Work 2008:
Hideaki Kawashima
Tomoriboshi

川島は手先の器用な学生で「オリガミ君」という愛称で呼ばれていた。実際に自分で使える照明器具を難燃紙で製作した。組み立てと解体が容易で部品交換が効き、更に強度が出るオリガミの手法（ノーメックス紙）を用いたらしい。私にはその詳細技法は解らなかったが、入り組んだ構造にすることで強弱をつけ光の破片を内外に散らす効果を生み出している。

素材｜半透明の難燃紙、2007年

Kawashima is good with his hands. He got a nickname of "Mr. Origami". He actually made lighting equipment for personal use using flame-resistant paper. He seemingly adopted the Origami Method (Nomex paper) that enabled easy assembly and dismantling and developed strength. I had no idea about detailed methods. The intricate structure offered varying strength and was effective for diffusing the pieces of light both internally and externally.

Material: Translucent flame-resistant paper

卒業制作 2004　上田夏子
Life Station

Graduation Work 2004: **Natsuko Ueda**

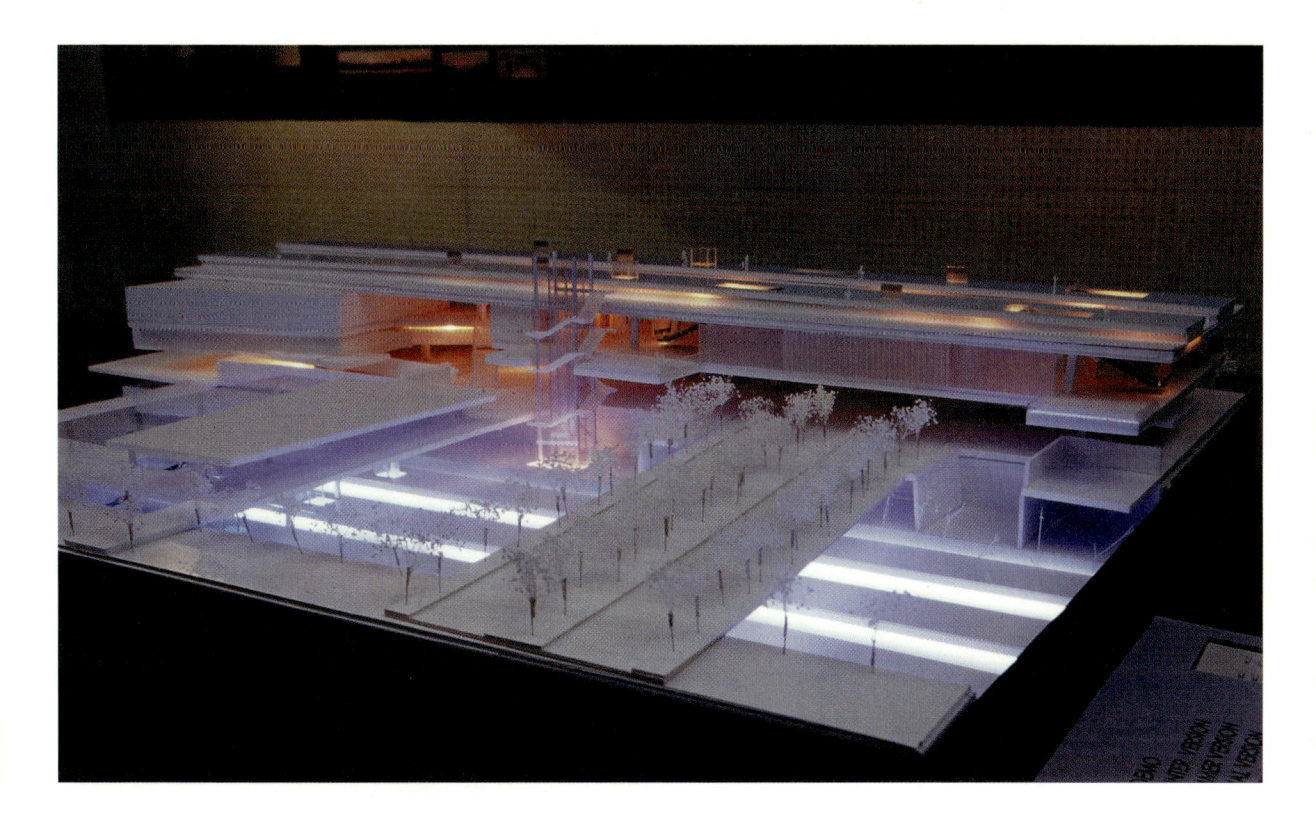

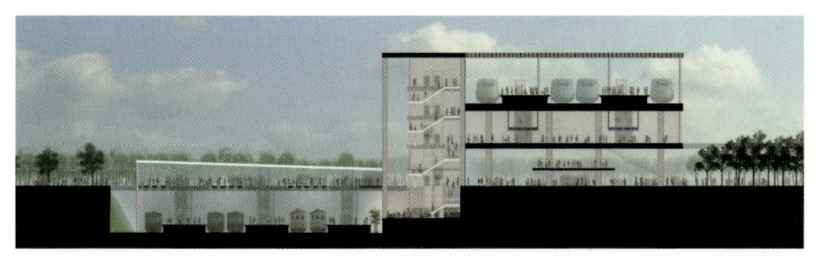

This was a plan to renovate Niigata Station. The student tried a fresh design to feel the breathing of users in the station square and on public walkways. It is highly appreciated as a good example of dynamics of architectural lighting design that is realized only by reflecting the seasonal changes and the night-and-day passage of time intended for light renewal, seen only in a snow country, and simultaneously proceeding with architectural and lighting plans.

Drawings and model containing light (joint project with Mio Fujita of the Department of Architecture)

新潟駅の駅舎のリニューアル計画である。駅前広場や公共的な通路空間に利用客の息づかいを感じさせることを意図して斬新な設計に挑戦している。光の衣替えを意図した雪国ならではの季節変化や、昼夜の時間の推移を照明計画に反映し、建築計画と照明計画が絡み合いながら、同時進行することによって初めて実現する建築照明デザインのダイナミクスを示す好例として評価される。

各種図面，光の入った模型
（建築学科・藤田未央との協同製作）

修了制作 2009　加賀見 鋭
光＝時のデザイン：CASE STUDY @ Ginza
Graduation(Master program) Work 2009: Kagami Satoki
Light or Time design: CASE STUDY at Ginza

快適な光の中でことさら大切な要素は「時間の流れ＝時のデザイン」である。光のデザインが自然なウツロイや時の演出を取り戻すことで、都市環境の光は新たな品質を手に入れることができる。

　光の適切な変化や動きによって夜の都市と人々の生活をもっと密接に関わらせる事は出来まいか。光環境を変化させることによって、人々の生活に個性的で楽しい都市生活シーンを与えたい。このプロジェクトは銀座4丁目のスクランブル交差点に光の変化を与えることにより快適で個性的な都市空間を創り出そうとする提案である。

A particularly important element of comfortable light is the passage of time, or time design. If the light design enables the restoration of natural lapse of time and of the dramatization of time, urban light obtains new quality.

　Isn't it possible to relate nighttime cities more to people's life through appropriate changes and movements of light? It is hoped that individualistic pleasant urban life is provided by varying light environment. This project proposes to create comfortable individualistic urban space by varying light at the scramble crossing at Ginza-yonchome.

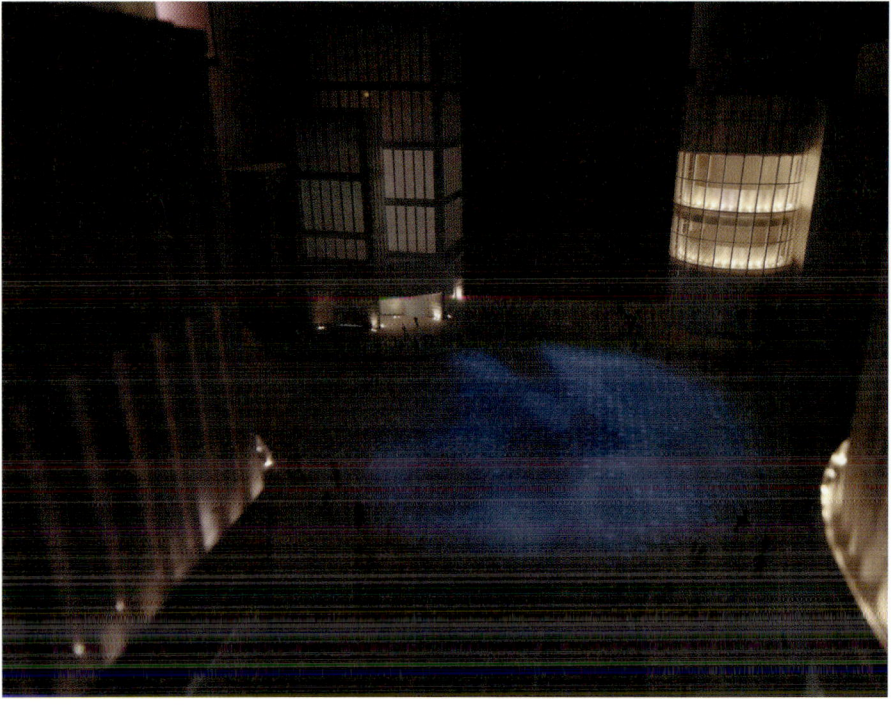

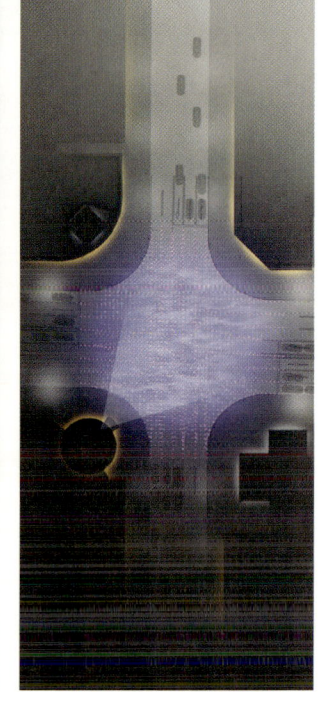

卒業制作 2007　池田俊一
Graduation Work 2007: Shun-ichi Ikeda

波紋
Ripples

霧に覆われ視界が朦朧とする薄暗い空間の中で聴こえてくる水滴の音と共に、静かに降り注ぐ水滴の波紋が霧を伝わって立体的に視覚化されることを意図している。長さ4メートルの水盤に落ちる水滴は波紋となりそれを天井に投影する。水盤に仕込んだ100粒のLEDはデジタル制御により20パターンの点滅サイクルを演出している。

　LEDと調光制御という先端技術を生かしながら水と霧という自然との融合を狙ったインスタレーションである。

素材｜水盤, フォグマシン, LEDランプ, 点滴ノズル, ほか

　The work was intended to offer the sound of dripping water heard in a dim space in haze and visualize three-dimensionally the ripples that were produced by gently falling drips and transferred through the fog. The drips falling in the four-meter-long basin formed ripples and were projected on the ceiling. One hundred LED particles built in the basin were digitally controlled and blinked in 20 different patterns.

　The installation was designed to integrate advanced technology represented by LED and dimming control, and nature in the form of water and fog.

Materials: Basin, fog machine, LED lamp, dripping nozzle, etc.

卒業制作 2005 窪田照彦
25年後の東京夜景・墨田川
Graduation Work 2005: Teruhiko Kubota
Nightscape in Tokyo in 25 years' time
along the Sumida River

作者コメント｜日本人にとって、灯りの文化といえる「提灯、行灯、障子越しの灯り」。それらは私達の心に原風景として残っているものであり、今も大切な財産だと思います。高度経済成長、24時間化、人々の生活スタイルが変わるにつれて照明も明るく、しかも眩しくなってゆきました。このままでいいのでしょうか。私達は、蛍光灯に群がる蛾ではありません。

面出評｜空間演出デザインの作品には、常に変貌する社会環境や、その揺れ動く価値観に対する斬新な解釈と表現が必要とされる。窪田の作品は表現の細部に渡って稚拙さを隠し切れない欠点はあるものの、私たちが抱える都市居住の現在的課題を解りやすく象徴的に指摘し、それに25年後の東京というフィルターをかけて実現可能な夢を語っている。不夜城のごときコンビニと隅田川にかかる橋の夜景。ここに現代の病魔があるといわんばかりに。

素材｜照明計画模型, コンセプト・パネル　映像4分

Comment of the student : The Japanese regard "the lights of paper lanterns, of lamps with a paper shade and through paper screen" as a culture of light. These types of light still remain in our mind as our original scenes and are precious properties. As people's life style has gone through changes such as high economic growth and around-the-clock life, lighting has become brighter and more glaring. Things should not remain that way. We are not moths drawn to fluorescent lamps.

Comment of Kaoru Mende : Designing space is said to require fresh interpretation and expression of ever changing social environments and wavering values. The work of Kubota pinpointed the current issues for city dwellers plainly and symbolically and talked about a feasible dream through a filter of Tokyo in 25 years' time although no poor expressions were concealed in detail. The nightscape featuring convenience stores and bridges over the Sumida River that resembled an all-night bustling area were presented as if to say that all the modern-day evils reside here.

Materials: Model of lighting plan, concept panel and video film (four minutes)

立面図 *Elevation*

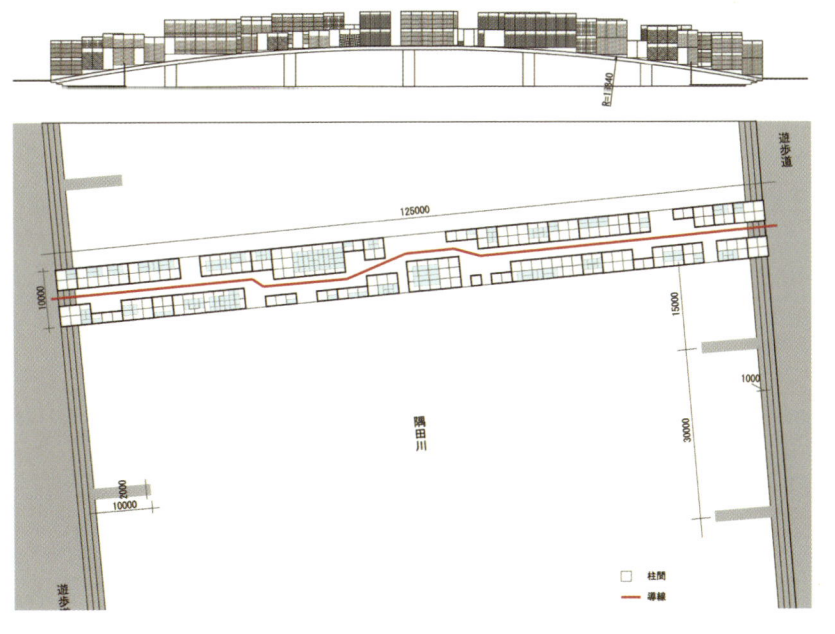

平面図 *Plan*

光を操る
MANIPULATING LIGHT

201

先生は現場を用意する　小池一子

空間演出デザイン学科のファッション専攻カリキュラムに、私は初年度（1987年）からプロジェクトとしての実習を導入し、また実施成果を社会人の目にさらすということを仕組んでいった。第一にチームワークの中で個人の創作の表現力を高めることを期し、「展覧会つくり」という形態をとるプロジェクトをスタートした。学生はどういう展覧会なのか、キュレーションのテーマを発想することから討議を始めなければならない。そして最終効果を上げるためには展示空間のデザイン、照明などが大きな役割を果たすことを理解する必要がある。

またファッションは服作りの技術的ノウハウだけではなく、社会・経済・芸術などの時代状況が創作の環境だということを知ることも大切だ。だが個別の制作密度と全員の展示の実際は、思うような成果をもたらさないのだった。私たちが試行錯誤をくり返していた「空デ」に、面出さんが着任された。

やがて私たちの夢は、二つの大きなプロジェクトにかなえられることになった。今は伝説となってしまったが、江東区佐賀町にあった食糧ビル全館の美術催事「EMOTIONAL SITE」で、2002年、そして2004年に武蔵野美術大学資料図書館展示室で行った「衣服の領域」展。これは私の退任に際しての催事でもあったので、レセプション空間の照明にいたるまで監修していただいた。アプローチに小さなキャンドルを使った導入が印象に残る。

食糧ビルは1927年竣工で、昭和初期の余裕にみちた空間を擁し中庭は廻廊に囲まれた欧風の広いパチオとなっている。3階に240㎡の柱のないホールがあり、そこを現代美術のギャラリーとして私の事務所が再生し、オルタナティブ・スペースとして活用していた。美術館でも画廊でもない美術現場の発想に共鳴してくれたギャラリストやアーティストが他の空き部屋も使うなどして1990年代の終わりには現代美術のメッカのような賑わいを見せたが、ビルの解体という事態を期にギャラリー活動も停止することになる。

面出ゼミはビル全館を使ってのクロージングのイベントに総力をあげて当たってくれた。屋上から光のシャワーをパチオにあてる試みで、面出ディレクターとゼミ生のやりとりが冬の冷えた空気の中に飛びかっていたことを思いだす。どのアーティストも今の日本、というより世界のアートシーンを先導しているような人たちだから、展覧会内容はずば抜けてよかった。学生のうちに、そのような展覧会の内側を体験できるのはすばらしいことだ。

それができる環境を教師たちが用意し、空間づくりをプロジェクトとして共に仕上げるという結果がそこで生まれていた。エモーショナル・サイトというタイトルの催事は美術界の出来事としてその年末の大きな話題をさらったが、エモーショナルの核心を演出したのは面出ゼミの光であった。
「衣服の領域」展は大学のキャンパス内で行われ、三宅一生さんの「プリーツ」を平面作品として壁面に展示したり、立花文穂の大量の印刷物を扱ったインスタレーションを設置したり、さまざまな作品への対応をしていただいた。

面出さんと私はこういった学内外の大催事でもあまり議論などしたことがない。あうんの呼吸というと妙に思われるかもしれないが、古い建物の再生を行っての"現代美術現場"とか服の領域を広げるため暴れるデザイン展とか、主題をさっと呑みこんでゼミ生と練って下さることに全面お任せという快感すらあった。面出ゼミからは、面出さんのそういう大きな器のような人材が出ると期待している。

小池一子　Kazuko Koike

クリエイティブディレクター／1936年東京都生まれ。1960年代にデザインに関する執筆、編集活動を開始。1976年、株式会社キチン設立。各種の美術展、展覧会などを手掛ける。1983年から2000年まで、日本初のオルタナティブ・スペース「佐賀町エキジビット・スペース」創設・主宰。現在は3331 Arts Chiyodaにて「佐賀町アーカイブ」運営。武蔵野美術大学名誉教授。くらしの良品研究所所長。

Creative director/Born in Tokyo in 1936. Started writing and editing concerning design in the 1960s. Founded Kichin in 1976 and organized art and other exhibitions. Created and presided over the Sagacho Exhibit Space, the first alternative space in Japan. Currently managing Sagacho Archives at 3331Arts Chiyoda. Professor Emerita, Musashino Art University, Director, Research Institute of Ryohin Keikaku.

Preparation of the Site by Teachers

In the first year (1987) I introduced into the fashion curriculum of the Department of Scenography, Display and Fashion Design the practice of training students through involvement in projects that were to be shown to the public. A project to organize an exhibition was initiated in the hopes that individuals would develop powers of creative expression within the context of teamwork. Students had to begin by discussing the kind of exhibition that would be held, that is, by suggesting possible themes for the exhibition. They had to understand that the design of the exhibition space and the lighting would play a major role in the effectiveness of the exhibition.

It was also important for them to know that fashion was more than just technical know-how about the making of apparel and that the social, economic and artistic conditions of the time form the creative environment for fashion. However, the work done by the students individually and the actual exhibitions organized by them were not as substantive as we had hoped. We at the Department were still trying to find the right approach to take through repeated trial and error when Mr. Mende took up his position on the faculty.

Eventually, two major projects that have since become legendary enabled us to achieve our objective: an art event entitled "Emotional Site," held in 2002 using the entire Shokuryo Building in Sagacho, Koto-ku, and the "Territory of Apparel" exhibition held in 2004 in the gallery of the Musashino Art University library. The latter was held on the occasion of my retirement, and Mr. Mende supervised all the lighting including that for the reception space. I remember vividly the small candles installed on the walkway to the space.

The Shokuryo Building, completed in 1927 in the early Showa period, had generous spaces. A cloister surrounded a broad, Western-style patio. My office restored a 240 square meter hall, which was free of columns, on the third floor and used it as an alternative exhibition space. Art dealers and artists who sympathized with the idea of a place for art that was neither an art museum nor a gallery used other vacant rooms in the building. At the end of the 1990s, the place was a lively mecca of contemporary art. However, the slated dismantling of the building put an end to such activities.

All members of the Mende seminar took part in organizing the closing event which made use of the entire building. I remember Mr. Mende and the seminar students trying to shower light onto the patio from the roof and shouting back and forth in the cold winter air. All the artists involved were people who are now at the forefront of, not just the Japanese, but the international art scene, so the content of the exhibition was outstanding. It was wonderful that the students were able to experience such an exhibition from the inside.

It set a precedent; henceforth teachers would prepare an environment that would make such an experience possible, and the creation of space would be a project completed through teamwork. The event called "Emotional Site" was a much talked about year-end event in the art world, and lighting by the Mende seminar was key to the staging of what I would call the event's emotional core.

The "Territory of Apparel" exhibition was held on the university campus, and Mr. Mende was called upon to deal with diverse works such as "Pleats" by Issey Miyake, displayed as a two-dimensional work on a wall, and the installation of an enormous volume of printed matter by Fumio Tachibana.

Mr. Mende and I did not engage in much discussion even for such large events inside or outside the university. To say that we were on the same wavelength may seem odd, but it felt good to leave everything to him. He would immediately grasp the theme, whether it was to create a site for contemporary art through the restoration of an old building or to stage a "disorderly and disruptive" design exhibition in order to broaden the scope of apparel, and develop ideas with seminar students. I hope that the Mende seminar will produce people of his high caliber.

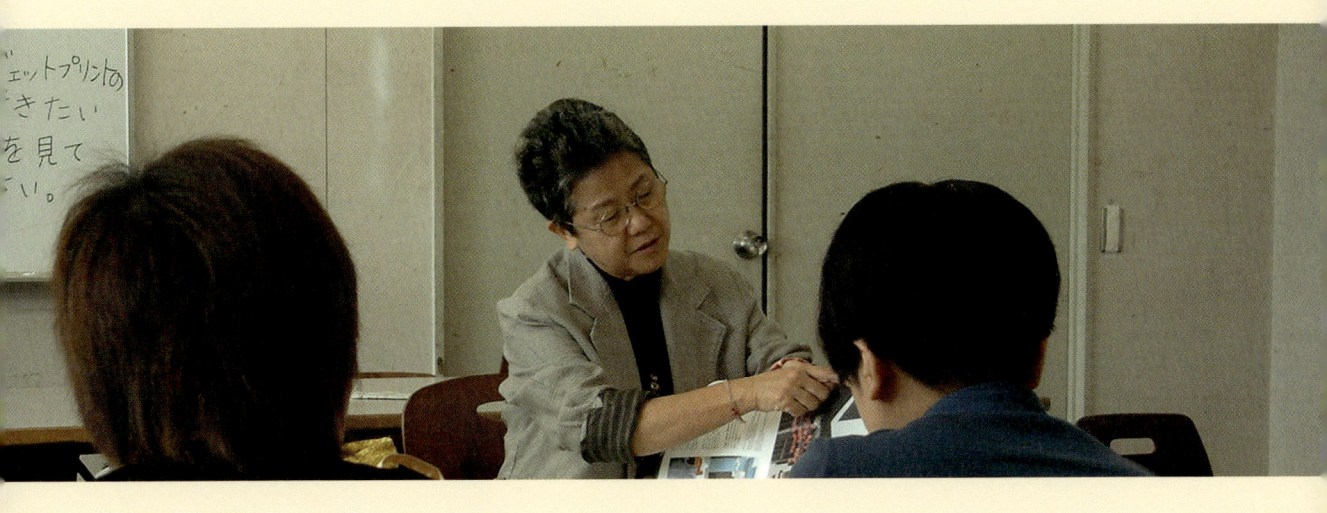

座談会1　卒業生がゼミで学んだこと

Discussion meeting 1 —— *What alumni learned from Mende seminar*

加藤直子　*Naoko Kato*（1期生）
池田俊一　*Syunichi Ikeda*（4期生）
六角望　*Nozomi Rokkaku*（4期生）
不殿晴子　*Haruko Fudono*（5期生）
中村将大　*Masahiro Nakamura*（5期生）

加藤　座談会日和とでもいうか、今日は面白い話を期待できそうですね（笑）。ここに集まってもらったのは、歴代の面出ゼミ卒業生のメンバーです。4期生から2人、5期生から2人、そして1期生の私の5人で、面出薫ゼミで過ごした時間や、そこから学んだ光やデザインとは何だったのか、そのあたりを中心に語りあえたらと思います。単に思い出を振り返るだけではなく、面出ゼミという教育環境を私たちなりにクリティックしながら、そこで得たものが今や今後にどうつながっているのか、そんなところまで見つけられたらとも思います。ちなみに面出ゼミには10期生までいて、総勢140人くらい。全員で語るのはさすがに難しそうですので、この場は今回の編集委員に集まっていただきました。ウェブ・クリエーターの六角さん、ライティング・デザイナーの池田君、デザイン学校講師の中村君、イベント企画で右に出るものはいない不殿さん、そして図らずも面出ゼミのお局になってしまった1期生の加藤です（笑）。本日、面出先生はこの場にはおられないので、このメンバーで思う存分本音で語っていただけるよう期待しています。

なぜ面出ゼミに入ったのか？
ゼミに期待したこと

加藤　今集まってもらったメンバーは、このゼミでやっている照明や光デザインの仕事を必ずしもしているわけではない。私自身もデザイン業界ですが、ランドスケープ分野に携わっているし…。でもきっと、今の仕事の発端は学生時代の頃からあったし、何かしらのものをつかみたくて面出ゼミに入りました。
中村　武蔵美では、大学3年生の後期にゼミに所属しますね。3年生後期から卒業までの約1年半を、所属ゼミで影響をたくさん受けていくような、そんな体制になっています。

写真左から｜六角望, 池田俊一, 加藤直子, 中村将大, 不殿晴子
Left to Right: Nozomi Rokkaku, Shunich Ikeda, Kato Naoko, Masahiro Nakamura, Haruko Fudono

「照明」ではなく、「光」のデザイン

池田 光のデザイン、その特徴から知ろう、ということですか？

加藤 うん、光のデザインというのは、建築とかグラフィックとかプロダクトとか、最終的に形になるデザインそのものではなくて、気配のデザインに近いのかな。

中村 環境や空気感のデザインかな。

加藤 形として出てこないぶん、誰かに伝えるのが難しそうに感じた。イメージとしてもっているけれど、それを図面に起こしたり、数字では表現しづらい分野かもしれない。

池田 ここにいるみんなは、そもそも何で照明デザインのゼミを選んだの？

中村 僕は編入組で3年生から武蔵美に入ったのですが、前いた美術大学が建築やインテリアのデザインをやっているところで、非常にポストモダンに熱心な学校でした。そういうところなので、形で勝負というか、デザイナーの作家性というところで、僕は個人的に悩んでいました。そんな大層な作家性があるのかとか、あるいはそれを外に出していくことに対しての気恥ずかしさや、恐れ多いと感じていた時期があって。形にとらわれたり、デザイナーの主観がもう少し出てこない方向でのデザインというのはないのかなと考え始めて、そんなときに偶然見ていたDVDに面出先生が出ていらした。「自分は形をつくるタイプのデザイナーではない。絵になかなか描けないような類いのデザインをやっている。あいつはデザインしたなと思われるデザインはしたくない」というようなことをおっしゃっていた。

池田 所属ゼミを決めるのに、3年生前期末に各ゼミの説明会があって、先生から学生に対してプレゼンみたいなものが行われる。でも、面出先生はすごく忙しいから説明会の当日に来なくて、あらかじめ制作されたムービーを見せられました（笑）。ムービーだけじゃ伝わらなさそうだけれど、面出先生のことは雑誌、テレビとか、実物の作品に触れる機会があるからほとんどの学生は知っていたと思います。

加藤 私の時期が初代ゼミだったけれど、先生のサインをもらうのに授業後に並ぶ学生がいたくらい。デザイナー面出薫が武蔵美に来る、というのは大学内でも大ニュースに感じられた。ただ個人的には、そもそも将来の職業選択の1つとして、ある種ニッチにも思えた照明デザインをやりたいわけではありませんでした。だから先生から何を学べるのか、悶々と考えていたと思います。照明デザインとは何だ？そんなところから始まり、やがてそもそも光のデザインって何だ？といった疑問にきつきました。

Why we chose Mende Seminar and what we expected

KATO: Today, five former members of the Mende Seminar are here to talk about what they did and learned at the seminar and criticize seminar education. In the ten years of its activity, the seminar has turned out about 140 members. First of all, would you tell us why you joined the seminar, what motivated you to do so and what you expected from it?

IKEDA: In the first semester every year, there is a seminar orientation for third-year students so that they can decide which seminars to join. Teachers make presentations of their seminars. But Prof. Mende was so busy he was not there on the orientation day, and there was a video movie presentation. Although there may be many things that can't be communicated through a movie, that was not a problem to me because I knew about Prof. Mende from magazines and TV and about his works.

I was very interested in what I vaguely regard as light in nature, such as Christmas illumination, a starry sky, thunderbolts, electronics and flames. It was when I was in avid search for such light that I came to know about the lighting designers Mikiko Ishii and Kaoru Mende.

KATO: I'm now working as a landscape designer, and I joined the light seminar because I wanted to do something different. It's not that I wanted to do lighting design, which in those days seemed to me to be a niche category. Instead, I was anxious to know what can be learned from Prof. Mende. I was always wondering what light design was.

NAKAMURA: I transferred to Musashino Art University as a junior. Back then, I was skeptical about form-oriented subjective designing. I happened to watch a DVD in which Prof. Mende was saying something like "I don't do design work that involves the creation of shapes. Instead, I produce designs that cannot be drawn or painted. I don't want to produce designs that make people think that somebody produced them." That impressed me tremendously. So, I really didn't care whether it concerned light design or not.

加藤　デザインする前の考え方みたいな?

中村　そうですね。こういう人がいると知り、武蔵美でゼミをもっているということだったので、それではいくしかないなと…。だから僕は、照明のデザイン、光のデザインのところからは入っていないと思いますね。学校の中での悩みをどうにか突破するというところからだったのです。

加藤　私は1期生だから、面出ゼミが何をやるかという前例がなかった。なので前例から決められなかったけれども、雑誌や実際の作品を見て、面出先生の作品にひきこまれた。言葉なしに、目で見てほんとうに美しかった。その風景に心が動いたことと、もう1つは、先生が団長を務める照明探偵団かな。まちの明かりを探したり、集めることを市民とやっている。見た目に美しい作品をつくりながら、日常の風景の見え方自体に疑問を投げかける活動をしているデザイナーにとても興味がわいた。

池田　僕は3年次に基礎デ*1から空デ*2に転科してきたのですが、基礎デは本命ではなかったので、入学の時点で数年後に転科する気持ちが固まっていました。本命は工デ*3という…。

一同　空デが第一希望じゃなかったんだ(笑)。

池田　みんなとは違うアプローチの仕方です。そもそも面出先生という存在は知らなかったけれど、漠然とクリスマスのイルミネーションだったり、星空だったり、雷、エレクトロニクス、炎とか、そういう自然の中の光というものにとても興味があった時期でした。一概に建築やアート、デザインというところには興味がなく、何とかして光を知りたくて、夢中になって探していたときに照明デザイナーの存在を知りました。石井幹子さんの作品集とかを毎日のように眺めていて、そんな中で面出薫というデザイナーのいることを知って、よく見たら武蔵美にいる。それで一度会ってみようと思いたち、1年生の頃に会いに行ったのがきっかけでした。照明デザインへのこだわりはなくて、あくまで自然の中の光に触れるのに、唯一教えてもらえる場所という期待で近づいた結果、今に至っています。

六角　私は、もともとは舞台とかTV関係の照明をやりたいと思って空デに入学したのですが、1年生の空デ論での面出先生の授業を受けて、高校生のときには知らなかった分野を見せてもらった。普段何も気にせずに歩いているまちの照明とか光とかをデザインしている人がいて、機能のためにデザインしているということを初めて知った。でも機能の前に、スライドを見ていて単純にきれいだなと思いました。光を照らすと、建物がこれだけきれいに見えるということを感じました。光をデザインするというより、光がアートのようにきれいに見えるということをそこで知って、面

KATO: I thought Prof. Mende's lighting designs were very beautiful. There was no need for words. It was simply beautiful to the eye. Another reason was the Lighting Detectives, a group headed by Prof. Mende. Its members are working with citizens to explore and sample urban lighting. I became very interested in him because he is a designer involved in activities to question the way townscapes look in daily life.

ROKKAKU: At first, I wanted to do theatrical or television lighting design. Then, I entered the university and attended Prof. Mende's classes and came to know that there are such things as architectural lighting and environmental lighting. Light can be expressed and appreciated artistically. That was interesting to me. That's why I decided to join the seminar.

FUDONO: Originally, I was interested in theatrical arts. I wanted to become a professional who can touch the hearts of people with a single object. But, after I attended various classes, it was light that my heart responded to most. I thought that making existing things look more beautiful rather than creating things from scratch suits me more.

Collecting light examples: Light Collection

KATO: Third-year students begin to work on seminar assignments in the second semester. The only common seminar assignment given every year in the last ten years was "Light Collection." Light Collection is divided into two categories: *natural light collection* and *urban light collection*. In collecting urban light examples, each student takes pictures of impressive light scenes and classifies the collected scenes. Let's criticize this assignment.

出ゼミに入りたいと思いました。

不殿　空デに入った頃は舞台美術に興味があった。その中でも、形よりも光に興味があり、とりわけ心動かされたのはライブ空間でした。そこでは無条件にテンションがあがる。デザイナーという職業名として語るよりも、1つのもので人の心を動かせるような仕事に就きたいというのはありました。ゼロからものをつくるより、すでにあるものに対して光によってさらに美しく見せるということのほうが自分にも向いている気がしました。建築でも舞台でも、いつも見ているものがお化粧をするように、よりきれいになるということに魅力を感じていました。

加藤　そうですか。みんな個人的な体験で感動したことを、どういうふうにデザイン経験として、あるいは学舎にして修得していくのかを考えていたのかな。そこが一部の人には面出薫と光でつながっていって入ったと…。

光を見つけることから始まる
課題｜ライトコレクション

加藤　私たちは、3年生後期からゼミの課題に取り組みますが、すべての代のゼミ生に唯一共通している課題がライトコレクションです。まちの中など日常的に目にする景色の中で、印象深い光のシーンを各自写真に撮って、それらを分類するというものです。4期生以降になると、分類するときに光の英雄、光の犯罪者に分けて、その理由を説明することをしています。英雄はつまり自分が良しとしたり感動したもの、犯罪者はなくしたほうがいいのではないかという批判的なもの。

池田　人によってどちらに分類するかは全然違って、ライトアップが好きな人はそれを英雄にしちゃうし…。

加藤　そうか、定義づけからも個人によるんですね。

六角　自動販売機が印象的で、半々くらいに別れました。犯罪者という人と、暗闇にある販売機は女の子の帰り道には英雄だ、というのがあって。

中村　美大という場所ではなかなかない機会なのかなと思う。自分がつくったものに対しての批評はするけれども、現状、それも日常の風景に対して評価することは初めてだった。何でそれがいいと思っているの？とか、他人の批評に対して自分もすぐには同意できなかったりするけれども、説明を聞いていると納得できたり…。ああいうものをストックさせて、共有していくというのは、面出ゼミの課題でも共通していたと思いますね。あの頃って、光の見方に慣れてなくて、結構ベタなやつを選んでいたかな。

不殿　私はベタベタだったね。犯罪者はパ

ALL:

* It's interesting to note that the same examples are called "heroes" by some students and "criminals" by others. Those who like lighting up classify them as heroes, and those who dislike it call them criminals. Vending machines are very interesting. Some say they are criminals because they wastefully emit glaring light, while others say vending machines in the darkness are heroes for girls on their way home. Opinions are nearly equally divided. Even though judgments are divided, you may think there is a point in what opponents say when you hear their explanations. Probably, those exchanges were important. The method of accumulating and sharing the criticisms thus obtained was also used in other assignments of the Mende Seminar.

* In a sense, there might have been some brain-washing by Prof. Mende. Maybe it was a time when some people began to dislike excessive brightness or excessive darkness.

* Some say fluorescent lamps glaring even at midnight are criminals, and others say a flame of a candle placed on a bar table is a hero because it creates a warm atmosphere. At first, we notice only extreme types of light. Working on this assignment, however, we gradually became capable of noticing more microscopic light rather than macroscopic light.

* I feel I became able to notice less visible, modest types of beauty such as light from a glass tumbler edge and droplets glistening on a leaf. It may have been the kind of assignment that helped us form the habit of carefully observing light.

Light event to redefine town–people relationship: Candle Night

チンコ屋で、英雄は雰囲気のいい露店とか。

池田　ある意味、面出先生の洗脳が入っているね。過剰に明るいとか過剰に暗いことに、嫌悪感が芽生えていた時期だったかもしれない。

加藤　始めは極端な光しか見えてこないんだよね。深夜でもギラギラ光る蛍光灯が、静けさを破るようで犯罪者だとか、バーのテーブルに置かれるろうそくの灯が情緒があって英雄にカテゴライズされるとか。でもこの課題を通して、徐々にマクロからミクロな光の存在に気づくようになっていく。

中村　ガラスコップのエッジの光だとか、植物の葉の上に光る水滴だとか、普段気づきにくい、ささやかな美しさを発見できるようになった気がする。

池田　そう、そして光を観察すること自体が習慣になっていくような、そんな課題だったかもしれない。

まちと人をつなぐ光のイベント
課題｜キャンドルナイト

加藤　その次になってくると、光を照らすことだけがデザインじゃないんだよ、という課題になってくるんですね。キャンドルナイトとか、闇をデザインすることで光をデザインするような課題に。私たちの代では、残念ながらキャンドルナイトはやっていないのですが、私たち1期生の代では、闇に対する意識とかそういう教育プログラムはまだなかったんです。4期生あたりから本格的に始動しているけれど、この課題はどうだった？

池田　もともとはカナダで始まった反原発のためのムーブメントであって、アメリカの原発政策に反対するために、電気を消してキャンドルだけで過ごそう、ということから始まりました。キャンドルだけでも十分過ごせる、そういうアプローチがキャンドルナイトの原点なんです。2003年に日本に入ってきたときに、当時の日本では今ほど原発が社会的な論点になっていた時期ではなかったので、どちらかというと新たなスローガンとして"電気を消してスローな夜を"というところで始まった。

加藤　コンセプト自体が変わってきている。

池田　そう、特に表参道のキャンドルナイトの場合は、総本山の竹村真一さん、面出先生がいて、協力してくれている佐藤卓さん、深澤直人さんに入っていただいて、わりとクリエイティブ色のあるイベントを展開していくわけです。一概に、キャンドルを置いてきれいだね、というところではなくなったのが、特徴的なところでしょう。表参道のときのように、まちを巻き込んだイベントを知識や経験のない学生たちがやるのはなかなか大変で、まず表参道がどういう場所か、というところから入る

KATO: Next comes an assignment that does not involve designing with light. Every year, a Candle Night assignment aiming to deepen the understanding of darkness was given mainly to third-year students. What did you think of that assignment?

IKEDA: The candle night scheme originated in Canada. It was an anti-nuclear campaign event designed to encourage people to turn off their lights and spend the night time with only candle light. When it came to Japan in 2003, it grew into a candle night event designed to involve a million people, and people were encouraged to "turn off the lights and take it slow." This assignment began because the organizers of the event included the cultural anthropologist Shin'ichi Takemura and Prof. Mende. Later, the graphic designer Taku Sato and the product designer Naoto Fukazawa joined the organizer group, enhancing the creative nature of the event.

ALL:
* The assignment involved an event to be held in Omotesando, a classy urban district, so everyone was making desperate effort. That's because coordination with the local community is a formidable task.

* It's a big event involving not only Musashino Art University but also other seven or eight universities and design schools. In our days, our seminar was playing a central role in organizing the event. So, we were busy going back and forth to coordinate with local shops.

のですが…。

中村 表参道を解剖して、どこにどんな店があるかとか、どこに暗闇があるかとか、どこで展示すると効果的かとか、そういったところから始まる。どちらかというとフィールドワークに近いのかな。

加藤 うん、フィールドワークをまちのイベントとしてどうやっていくか、という。

池田 そう、そこに"電気を消してスローな夜を"というのを突き詰めて発見することだったかな。明るさとか暗さを調整することで、普段は気づかない暗闇の美しさや落着き、きれいさを知るきっかけにはなった、そんな気がする。

不殿 私たちのときは、イベント運営の中心的立場にゼミがあったので、参加カフェの数を増やした時期で、お店にお願いにたくさんまわっていて。

中村 不殿さんと企画書もって2人で行きましたね。

不殿 そうそう、お店に行くと「毎年楽しみにしているんです」といってくださる人がいたり。イベント前になると、表参道の店ではキャンドルの準備をしていたり。明るすぎる現状に気づいて、キャンドルで過ごす暗さの魅力が浸透している実感があったよね。

六角 表参道という、とても活気があって長い歴史のある場所で、イベントを企画できたことが貴重だった。企画説明やアイデアをまちの人に伝えたり、店の人から了解を得るプロセスが大変だったし、そんなさりげないことに1つ1つ時間がかかった。そんな集大成として、イベント当日はほんとうに嬉しかった。

中村 「きれいなものをつくる=デザインすること」ではない、フィールドワークを通してそう強く感じました。形をつくることだけではなくて、明りを灯す行為そのものをまちの人たちと一緒に考えることもデザインですね。

さっと照らして、さっと逃げろ！
課題|ライトアップ・ゲリラ

加藤 フィールドワークが多いゼミ、というのが特徴なのですが、ライトアップ・ゲリラというのは、ゲリラっていうくらいだから、まちの人に何も知らせないで実行するんですよね。

中村 でも結構、用意周到にやっていますよ（笑）。

不殿 それも、お願いしてまわったりしたよね。

中村 ぼくらのときは、農家のビニルハウス、あと玉川上水、お寺、立川清掃工場の煙突でしたね。

加藤 フィールドワークといっても、逆のアプローチだね。キャンドルナイトは照明のあふれているところに、どういうふうに

*It was an invaluable experience because we were able to plan an event in Ometesando, a very lively, time-honored area. Explaining the plan and idea to the local people and obtaining the consent of the shops was no easy task, and each of those preparatory tasks was time-consuming. I was therefore really happy on the day of the event because it was the culmination of those efforts.

*When I saw the reaction of people in town, I realized the magnitude and possibilities of what design can do. Creating forms is not the only way of designing. Thinking about the act of lighting together with townspeople is also a way of designing, isn't it?

Light up and run: Light-Up Guerrilla

KATO: One characteristic of the seminar is that there is a lot of fieldwork. There's an assignment called "Light-Up Guerrilla." Since it's called "Guerrilla," is it carried out without notifying townspeople in advance? I hear that the key to success is to light up quickly and run away.

ALL:
*Despite its name, actually we did a considerable amount of advance coordination.

*It was another fieldwork assignment, but the approach was the opposite of the Candle Night approach. The aim of the Candle Night project was to create darkness in a brightly lit space, while the aim of the Light-Up Guerrilla project is to provide light in a lightless space. In some cases, it began by creating a dark space. The seminar members were the only spectators. It was an event not designed to attract spectators.

闇をつくるか、というところだけれど、ライトアップ・ゲリラはそれとは対照的に、光のなかったところに光を当てるとどうなるか、ということをやったんだね。どうでした？　集客性のないイベントをやるというのは。

中村　そうですね。見ているのはゼミ生だけですね。

不殿　"パッと照らして、パッと逃げろ"、みたいなキャッチコピーだったもんね（笑）。

中村　結構もたついていたけれどね。

池田　あれって何だったの？

加藤　光を照らすことのメリットもあって、それはきっと純粋に光の表現ができることも1つだよね。どうしてもキャンドルナイトって啓蒙的な部分が含まれます。だけど、表現として光をバンバンつくってよいというのは、もちろんあるわけで…。

中村　確かにそうですね、かなり表現に特化していた。

不殿　普段照らされていないものに光を当てて、どう表情が変化するかとか、実験的なところが多くて、なおかつ当日まではやらないので、自分たちなりの予想を基に場所を選んで計画を立てる。それで、実際にやってみて「全然違うじゃん！」となったりとか（笑）。

加藤　失敗もあるんですね。

不殿　そう！「やっぱりきれいだったね」とか、予想以上にきれいなパターンもあって…。

中村　玉川上水とか、絶対きれいだと思っていたら結構こけました。一方で、ビニルハウスとか煙突とか、うまくいかないんじゃないかと思っていたのがきれいだったりとか。

不殿　そう、普段の光の体験だと、やっぱり自然がきれいだと感じやすかったりするのだけれど、たとえば街路樹に当たる光とか。でも、あのときは使っていた照明器具がビームとか、LEDとかだったからかもしれないけれど、自然のものを照らすのはこんなに難しいんだという実感はあった。自然のもののほうがきれいに見せやすいと思っていたから。「やられた〜！」というのはあったね。

加藤　そうか。この課題も1期生はやっていない…。

中村　ああ、ライトアップ・ゲリラ楽しかったですねぇ！（笑）。

加藤　フィールドワークが重要という流れに入ってきている、ということはあると思うんですよ。雑誌とかTVで見て「ああ、きれい」と思うものを再現しようとしてもそうならない、そのギャップが光のデザインなんだよね。ちょっとカメラの設定変えただけでも全然違う光に見えるじゃない？　そういうイメージとか、すごく共有しづらい中でフィールドワークのやりがいがあるのかもしれない。一方で、フィールドワークのデメリットというのもあったと

* It was a fieldwork assignment mostly focusing on expression.
It was experimental in many ways. Although we select a site and draw up a plan on the basis of our estimation, we may end up being dumbfounded by a totally unexpected result. In other cases, the result may be more beautiful than expected.

* I realized that illuminating natural objects is so difficult.

* Ah, I really enjoyed Light-Up Guerrilla! (grin)

KATO: So, Prof. Mende really imprinted the importance of fieldwork on us. On the other hand, didn't you feel that scholarly research time was too short?

ALL:
* Yes, that may be right. I think the idea is to give priority to fieldwork rather than scholarly research, and students are supposed to do scholarly research by themselves. I feel that I spent more time outside the seminar room.

* We were doing practical training again and again rather than acquiring knowledge or learning theories.
It was like trying to make you body remember. Really, we were trying to feel and absorb.

Ten years consisting of three periods

KATO: Having talked with former seminar members of different generations, I have realized anew that those of us in the first batch of seminar students did not do such assignments. It seems that many of Mende Seminar's high-profile classes were conducted during the fifth and sixth years. This means that the period until around the fourth year was the "sprouting period" of the Mende Seminar. Looking back on the last ten years, I have now realized this.

思う。知識をインプットする時間が圧倒的に欠けてしまう。
中村　確かにそれはありますね。
池田　それはこのゼミの特徴なのかもしれない。現場主義というか、やってなんぼという。
中村　確かに、ゼミ室以外の時間のほうが長かった気がします。
池田　圧倒的に長かった…。
加藤　ある知識体系や理論よりも、実践的なトレーニングを繰り返していた。
中村　体にたたきこめ！みたいな。
不殿　ほんと、体感して吸収していたね。

面出ゼミの「3つの時代」と時代をまたがるルール

加藤　何だかこう話していると、私のいた1期生は改めてそういった課題をしてきていないんだなあと…（苦笑）。面出ゼミの看板となる授業が、5〜6期生頃にもっとも多く展開されている。つまり、4期生頃までは面出ゼミの「発芽期」だったと。10年間の歴史を振り返ると見えてくるものがありますね。
池田　面出先生が試行錯誤しながら課題を展開されているのが見てとれますね。
加藤　つまりは、3期生くらいまでは授業の実験期間。この書籍には載らないような課題を数々やってきた世代です。そういっ
た小さな種まきが数年間あり、それらが4期生以降で完成していった感じ。今ここに集まっている4〜5期生は、ゼミの看板授業が対外的にも知られるほどになった「発展期」でしょうね。
中村　その後の6期生以降は？
不殿　「成熟期」なのでしょう。
池田　成熟期には、各課題の方法論も形になってきて、学生もゼミの教育意図をあらかじめつかみやすかったのではないでしょうか。課題だけではなくて、チームワークの作り方も、先輩からの経験を聞いたりしてそれなりにゼミの伝統が見えてきました。
加藤　そして各代に共通していえるのが、「面出ゼミのルール」です。
中村　そう、ゼミの一番始めの授業で配られた紙に"メールは1日に2度はチェックするように"と書いてありましたね。
加藤　連絡はこまめにとか、事前に準備してこいとか、まるで企業研修の1年目のような。
不殿　こういう課題をやります、というリストの下に書いてあったね！
中村　面出ゼミの3原則が、"美しい挨拶"と"環境美化"と"遅刻をしない"だったかな。
不殿　環境美化なんてあった？
六角　あった、あった。
中村　だって結構ほら、片付いていないと先生が嫌がっていたじゃない。課題の進行

ALL:
* I can see that Prof. Mende developed various assignments by trial and error while working as a busy designer.

* That is, the period until around the third year may be called the class experimentation period, in which various assignments that are not mentioned in this book were given. It seems that after those years of "seeding," assignments were gradually refined and completed during and after the fourth year. The students in the fifth and sixth years who are here today spent their student years during the "developing period," in which the high-profile classes of the seminar became widely known outside the university.

* During the maturing period from the seventh to the tenth year, the Mende Seminar began to assume characteristic expressions as the students not only tackled the assignments but also learned from advice from older students and graduates concerning, for example, how to build teamwork.

* Common to all generations were the "Mende Seminar rules." A sheet of paper that was handed out in the first seminar class said "Check your mailbox at least twice a day." The rules also required keeping in touch with one another, making preparation in advance and stuff like that. It was like on-the-job training.

* If I remember correctly, the Three Rules of the Mende Seminar were Greet Beautifully, Beautify the Environment and Be Punctual.

* The seminar room was messy while an assignment was in progress, but we tidied up the room before its presentation. We were highly motivated to make a good presentation. To do so, we assigned a time keeper and measured the presentation time.

中はゼミ室をごちゃごちゃにしていたけれど、プレゼンの前にはみんなきれいにしていた。

不殿 照明模型が出るときは、ちゃんとブルーフィルタを蛍光灯に巻いてとかしていたしね。場に臨む姿勢というか。

池田 プレゼン意識は高かったですね。

不殿 そうだね、プレゼンはすごく重んじていたよね。

池田 自分たちでタイムキーパーとか決めて時間計っていたし…。

加藤 プレゼン前のセッティングしっかり、普段の生活習慣には厳しめだったなあ。

世界各地の光に触れる

六角 チームワークができてくると、課題のスピードはぐんぐんあがる。ひと学年でこなす課題量がどんどん増えていった印象がある。

加藤 国内におさまらず、次は世界照明探偵団!なんて、課題は日本に限らない。どんどんスケールが大きくなっていって…。

中村 ほんとうにいろいろなところに行っていますよね。

不殿 ほんとうにね、身分不相応というか、贅沢なゼミだよね。

中村 何を批判するかっていったら、そこなのかな。

加藤 学生の出費が多いという(笑)。デンマークに、北京、シンガポール、場所を変えてのフィールドワークという目的もあったし、現地のデザイナーに会ったり、ショールームの見学もしているんでしょう?

池田 面出さんが団長である世界照明探偵団で、各国のコアメンバーがいて、その人たちのホームタウンを訪れてフィールドワークを行うのに、ゼミ生はお手伝いとして参加しました。

加藤 具体的に手伝いというのは?

池田 現地でのワークショップに参加する。ワークショップの企画自体は、僕らの代はシンガポールで、シンガポールの4〜5つの地区にチーム分けをして、その地区に対して2時間くらい調査をして、あとでまとめたものをプレゼンテーションした。それは多分、今も共通だと思う。

六角 日本でも、テーマに沿った光を調査してから現地に行ったよね。

中村 面出さんがフォーラムのプレゼンテーションで使う資料の撮影とかね。それこそ東京の夜景を象徴する場所として渋谷、新宿の夜景などを撮影しに行きました。それもまたフィールドワークになっていましたが…。

池田 世界照明探偵団は、現地の人と一緒になって光のディスカッションをしながら、一緒に1つにまとめて発表するという場になっていた。それを世界各地でやっているわけだから、国の違いとか系譜的に並

*Our lifestyle habits were relatively strictly controlled.

Experiencing light in foreign countries

ALL:
* As teamwork improved, assignment fulfillment became faster and faster so we became able to make trips to other countries every year to participate in the forums of the Transnational Lighting Detectives.

* In each country, there are core members of the Transnational Lighting Detectives, which is headed by Prof. Mende. So, we visit the chapters in their countries to do fieldwork.

* We visited Singapore. The city was zoned into four or five areas, and different teams studied different areas. Later, we made a presentation of the results thus obtained. Probably, that has not changed yet.

We can discuss light with local people and put together the results and are then given an opportunity to make a presentation. That's an experience that can only be had by Mende Seminar members.

* After returning to Japan from an overseas seminar trip, we notice we feel more strongly than ever that we are Japanese. Prof. Mende says we should rediscover In Praise of Shadows. Our way of looking at things has become deeper at the DNA level. In fact, I feel that various seeds were sown in those days.

* In the overseas activities of the seminar, we were able not only to see light in other countries but also to meet people with different values and realize how different they are from ours.

べると認識できますね。面出ゼミならではの体験かな。

中村 海外から行って帰ってきたときに、より自分の中の日本的な認識が強くなったかな、というのはあります。海外のデザインってずっと格好いいと思っていたけれど、ヤコブセンの現物とか見て、北欧だから出てくるものなんだなと感じた。自分たちの体からは、やはり違うものが出てくるのだろうと思いました。面出先生も陰影礼賛の再発見というようにおっしゃっているけれど、自分たちのDNAレベル、たとえば夏の縁側が気持ちいいとか、春の木漏れ日がきれいというような感性は自分たちのオリジナルだから、もっと大事にしていいと思うし、むしろ育てていかないといけないかなと思った部分です。どうしても日本のデザイン教育はモダニズムが根底にあって、そこからスタートしていくけれど、いくら国際様式とはいえ、自分たちが完全にその中に同化することはできないということは、探偵団に参加して気づいたことでした。反面、面出ゼミはフィールドワークが多く、そういう理論教育を突っ込まないこともありました。そこができると自分の直感が言葉で整理されて、より納得のいくものになったのかもしれません。

加藤 日本の文化はこういう特徴があります、なんてテキストベースで読んだりするけれど、自分の感性でとらえないときっと実感としてはもてないんだよね。頭ではこういうことだとか、こういうのが美しいというのを覚えたとしても、それが伝えられるかとか、ほんとうにそれがそうなのかと疑うところからというのは、自身の体験からしかきっと出てこない。そういう批判的な視線というのは、こういうフィールドワークを通じて与えられてきたけれども、それをもう1回持ち寄ってディスカッションする機会は、意外と少なかったのかもしれません。

中村 実はあの頃に、いろいろな種が植えつけられている感じもする。

加藤 発芽はそのときじゃなかった？（笑）

中村 発芽はない。むしろ卒業後だったかもしれないです。

池田 いろいろな人のいろいろな意見を聞けたというのは、自分との温度差を知るにはいい機会だったかもしれません。

中村 美大生はどうしても我が強いとかがあったりするけれど、いろいろな人の話を聞くことで、我を崩して行くようなことがあったと思います。

池田 日本人だけと話しているときと、海外の人と話しているのとでは、そもそもの価値観が違うから、いいと思うものに共感を得られなかったりとか、そういうことを初めて体感しました。

Professor Mende as practicing designer

* One of the advantages of art college education is that students can learn from practicing designers. In that sense, the Mende Seminar was really where we were able to learn by looking at Prof. Mende working as a designer. He was more a real designer rather than a university professor.

* Prof. Mende was a very busy designer, and everything taking place around him was fast-paced. "Oh, a real designer is this busy," I thought, and I was really surprised.

* I thought Mende Seminar members shared a very strong sense of professionalism. Art college students tend to be conceited because they are artists, but we were free from such conceit. We were making great effort to discover ways, for example, to communicate or express our thoughts effectively.

* It's not that we were expected to come up with great ideas. Instead, we were always being trained to convey what may be called our common sensations effectively. Moon light is beautiful, and sunshine filtering through leaves makes you feel at peace. It's about how we can inspire such common sensations in what situations.

* Experience with the Mende Seminar makes us more sensitive to light as a design medium. We gradually acquire the habit of always responding to light encountered in everyday life such as natural light and street light.

Process before result: Mende style

* By the way, Prof. Mende's assignment grading criteria were a mystery.

実務者としての面出先生

池田　美大教育のいいところは、第一線のデザイナーが大学で課題を受け持つ、ということですかね。

加藤　社会に出た後の実務とつながったデザイン教育が受けられるのかもしれない。興味を抱く実務者（デザイナー）と近い距離で同じ空気を吸ってみたい、と望んでいました。

中村　つまりは見ておぼえろとか、俺の背中についてこい、という。

加藤　背中を見ておぼえる、ほんとうにそういう世界で、面出先生は大学の先生というよりは生身のデザイナーでした。

池田　まさにデザイナーでしたね。

加藤　学生も大学とLPA*4を行き来するという機会が非常に多かったと思います。

不殿　先生から流れ出す空気はとても早い。それは面出先生が実務で超多忙を極めるがゆえにね。

六角　ああ、こんなに忙しいものなのだ、ということだけでも肌で感じていた。

中村　一方で、今僕は教育の場にいるのですが、自分の専門があやふやなんで何やろうかなというときに、結構今の学校では学生にアイデアが出ないとか、展開力が弱くて悩んでいることが頻繁にあって。そうやって、自分の中でのデータベースがないということに気づいたんですね。デザイン演習というのがあるのですが、世の中のいいデザインと悪いデザインを50例ずつ集めてこいと学生にいって、まさにライトコレクションの手法なんですけれど、その中でちゃんと説明させていくんです。そうすると、学生の初期の手が動かないというのがなくなってきました。体験を通じるというのは必要なのかなと思います。

加藤　そうですね、理論と実践というのはしかるべきバランスがあるけれど、それがどうしても偏ったりするし、面出ゼミでは理論とか体系づけというのは自己学習に委ねられた部分が非常に多かったです。多かったけれど難しいよね。デザインというのは感性を育てる、そっちを20代前半では必要とされるという風潮もあると思います。そこを徹底的にやってくれたのかなという気はしています。時間を守るとか、体裁を守ることとか、ほかの人から見れば特別な環境に映る美術大学の中も、実は受けていた教育というのは非常に基本的で一般的なことだったと思います。

池田　面出ゼミは、ほかのゼミに比べてプロ意識がすごく高かったですね。自分はアーティストだ、という美大生にありがちなうぬぼれがなかったというか。いかに考えていることを伝えるかとか、吐き出せるかとか、もらいにいかに掘く労力が大きかったです。

加藤　ものすごいアイデアを求められて

* I once asked the professor about it. He said that the process was important. Probably that means the period of trials and errors is important.

* That's really like Prof. Mende. He's always saying that a sketch filled with memories is better than a beautiful photograph. He seems to mean that because we are not artists, the real value lies in how you struggled.

DNA of Mende Seminar carried on?

KATO: Now the Mende Seminar is ending its history. How do you think Prof. Mende's way of thinking or genes will be carried on?

ALL:
* Many former seminar members have been working in the lighting industry, so the DNA of Prof. Mende will be carried on for a long time to come. There are also a substantial number of graduates who have entered other industries such as architectural and interior design. But Prof. Mende did not say he wanted us to go into the lighting industry. So, am I wrong to think that he left us something different other than the genes for contributing to the lighting industry?

いたわけではなく、誰もがもっている感覚みたいなものを上手に伝える訓練をずっとやっていたということですよね。月の明かりはきれいだし、木漏れ日に心が安らぐし、そういう当たり前にもっている感覚をどういうシチュエーションでどういうふうに引き出していくか。

不殿 それにプレゼンとグループワークは鍛えられたよね。

中村 実社会に出たら当たり前なのかもしれませんが。ところで、空デの中で面出ゼミは特殊な雰囲気でした？

六角 舞台美術とかディスプレイの専攻も含む空デの中で、面出ゼミでは光という集中した素材に対して過敏に反応する。だから、ほかとは違ったオタクっぽい雰囲気というか、面出ゼミ生と歩いていると、だいたい無意識に「あそこの店の照明は明るい」とか「あそこの照明器具が切れている」とか、そういうことをいっているのを聞くから、自分の生活の中で自然光に対しても、まちの明かりに対しても、反応するというのが自然に身についている。

中村 美大にいるとデザインとは特別なことのように思いがちだけれど、日常の暮らしと密着しているというのが、面出ゼミでわかったことです。その暮らしとか生活文化が国民性につながっていたり、デザインを通じて、何のためにデザインするのかとか、デザインの先に何があるのかを見据

えるパースペクティブというのは、面出ゼミにはあったかもしれません。

結果よりもプロセス主義が面出流

中村 ところで、面出先生の課題採点の評価基準は謎だったというか…。

不殿 謎だった！

中村 僕は職業柄、それをつねにやっているわけだけれど、面出先生は何を基準に点をつけているのかと思いまして。グループ課題の評価はすごく悩むところで、機械的に出欠でつけているのか、面出ゼミではプロジェクトで動くから、単一の課題の評価というわけでもなさそうだし。僕の場合はいつも同じ評価だったから、適当につけているんじゃないかと思っています（笑）。

不殿 自分で自分の課題に納得がいっていないことが多かったから、もらった評価が思っているより良いことのほうが多くて。それはモチベーションを保つためにそうしてくれたのか、それが不思議で一度質問したけれど、そのときはプロセスが大事という話をされた。試行錯誤した時間が大事なのか、いいものができてもポンと簡単にたまたま出てきたものならそれはどうなんだっていうか…。

中村 プロセス重視なのかな。

池田 それは面出先生にはあるよね。きれいな写真よりも、想いの詰まったスケッチ

* We have been greatly influenced by his attitude and philosophy, not by his skills or expertise alone.

* I think he may have taught us how to discover a connection between society and us. By society I mean things occurring around us. He may have taught us how to propose such a connection in the form of a design.

* The attitude of questioning the society we live in may be a great gene.

* Needless to say, it's necessary to learn figurative art skills, but I think an important thing I learned is the attitude to criticize whatever is taking place in everyday life or before our eyes.

* The attitude to question whatever is occurring around us or question ourselves with the power of design and tight and unique teamwork— these may constitute the DNA of the Mende Seminar.

のほうがいいっていわれるし。

中村 今思い出した言葉で、"どんなスケッチでもとっておきなさい"といわれた。自分のちょっと考えて書いたものとかは、絶対捨てないとかいわれていました。

加藤 何で結果論じゃなくてプロセスなんだろう？

池田 デザイナーだから、どれだけ何に困って、何に苦戦したか、というところに価値があるのかもしれません。

加藤 アートじゃないということでしょうか。

池田 うん。これだけ大変な思いをしている、というのを残すという。

中村 ライトコレクションにしても、経験値を残すやり方について、いつも気にされていたのかもしれません。思ったことや感じたことは、その瞬間にはすごく印象に残っているけれども、その後体に残るかというとなかなか難しくて。面出先生のレジュメの中に"大げさに感動することが大事"と書いてありました。そうしたことで、自分の中で感じたことが消化され、それをアウトプットできるようになっていくのでしょう。あとは、何かをやったら必ずまとめるというプロセスは面出ゼミでずっとやっていました。ボリューム多いんですけれどね。

面出ゼミのDNAは残るのか？

加藤 面出ゼミが幕を閉じる今、先生の思想みたいなものがどこにあって、どう残っていくのでしょうか？今話していたように、デザインの技術そのものではなくて、日々の暮らしの見方を変えるようなところが多いよね。その具体例を示すような先生の台詞をもう少し思い出してほしい。たとえば、恋愛でもいいし、食事でもいいし、全然違うところでもいいのですが。

池田 昔いわれたのか、最近いわれたのか憶えていないけれど、「トイレットペーパーを換えられないやつはデザイナーをやめろ」って（一同大笑）。交換できる思いやりのないやつはだめだ、と。

中村 それとはちょっと違うけれど、新正卓展の企画するという課題の初回の授業で、「図面とか必要資料は揃ってないの？」といわれて、「これからです」という話をしたら、「今日デザインやっちゃうんだよ」といわれて。要するに、みんなは初回はオリエンテーションのつもりでいたら、面出先生の中ではもうプランつくっちゃおうと思われていた。

加藤 スピード的な話？

中村 そうですね。

池田 ゼミは面出先生の普段の仕事の仕方ですよね。

中村 ゼミの授業で思ったのは、ブレン・ストーミングをよくやっていて、思いついたことはどんどんいっていいということ。美大生の心境としては、アイデアが出るまで黙っておこうとか、いってかましたろうとかあるんだけれど、ノーマルな意見でも発展させていくことで、ちゃんとデザインは洗練されるというプロセスがあったと思う。トイレットペーパーの話にもつながるかもしれないけれど、当たり前のことに対してちゃんと気配りができるという。

不殿 そうそう！ご飯食べに行っても、特に男の子だったように思うけれど、「○○ちゃんに醤油とってあげなさいよ」とか、気配りという点、生活習慣というところの指摘は結構あった気がします。

中村 「ホラ、はるちゃんのお酒がなくなってるじゃない」ってね。

不殿 で、面出ゼミの遺伝子ってどういうところなのだろう。

池田 照明業界では根強く残るけれどね、それ以外はどうなんだろう。

加藤 実際、照明業界にいるのは卒業生の何パーセントくらいかしら。

中村 各代で1〜2名？多いときは3名くらいいるのかな。だいたい10パーセント強。比較的新しい卒業生は、照明業界に進む印象が強いですか

池田 いますね、照明メーカーとか。

中村 照明じゃなかったら、けっこう建築

分野も多いですよね。

池田　インテリア事務所とか、進学ももちろんいます。

中村　何だかんだデザイン関係か、照明業界に残っている面出遺伝子というのは、ライティング技術とか、考え方、捉え方のところだろうけれど、でも学生の中にはそれとはまた違うところで遺伝子が残っているのかなという感じもしますね。

不殿　うん。面出先生は照明業界に必ず就職してほしいとはいっていなくて、でも遺伝子というのはどういうことか、ということだよね。

中村　そう、照明以外の遺伝子、照明も含めたプラスαのことを考えているんだろうな。

不殿　先生のおっしゃっていた言葉よりも、先生の姿勢とか哲学とか、そういうことに影響を受けていることのほうが多いから。技術というわけでは決してないもんね。

池田　技術なんて教えてくれない。

加藤　面出先生は、社会と自分との接点の見つけ方を教えてくれたのかも…。社会というのは、自分の身の回りで起きていることね。そことデザインでつくれる接点を。

中村　社会に対して疑問に思うこと、そうしたことを光やデザインを使ってどのようにアピールしていくのか。それは光を見つけたり、つくったり、個人課題やフィールドワークを通してたくさんの光の経験をしました。

池田　どんな光をどうやってつくるかという、造形的な力をつけることはもちろん必要だと思います。でもそれとともに日々の生活の中、あるいは目の前で起こっていることをクリティックしていく姿勢を身につけられました。

加藤　デザインの力によって周囲のことや自分を問う姿勢、そして強固でユニークなチームワークこそが、面出ゼミのDNAかもしれません。

2012年6月24日　新宿御苑にて

*1　基礎デ｜基礎デザイン学科
*2　空デ｜空間演出デザイン学科
*3　工デ｜工芸工業デザイン学科
*4　LPA｜面出の主宰する照明デザイン会社
　　　　　Lighting Planners Associates

座談会2　ゼミで伝えたかったこと

Discussion meeting 2 —— What we tried conveying to students

面出薫　*Kaoru Mende*
林三雄　*Mitsuo Hayashi*
岩井達弥　*Tatsuya Iwai*
角田真祐子　*Mayuko Tsunoda*

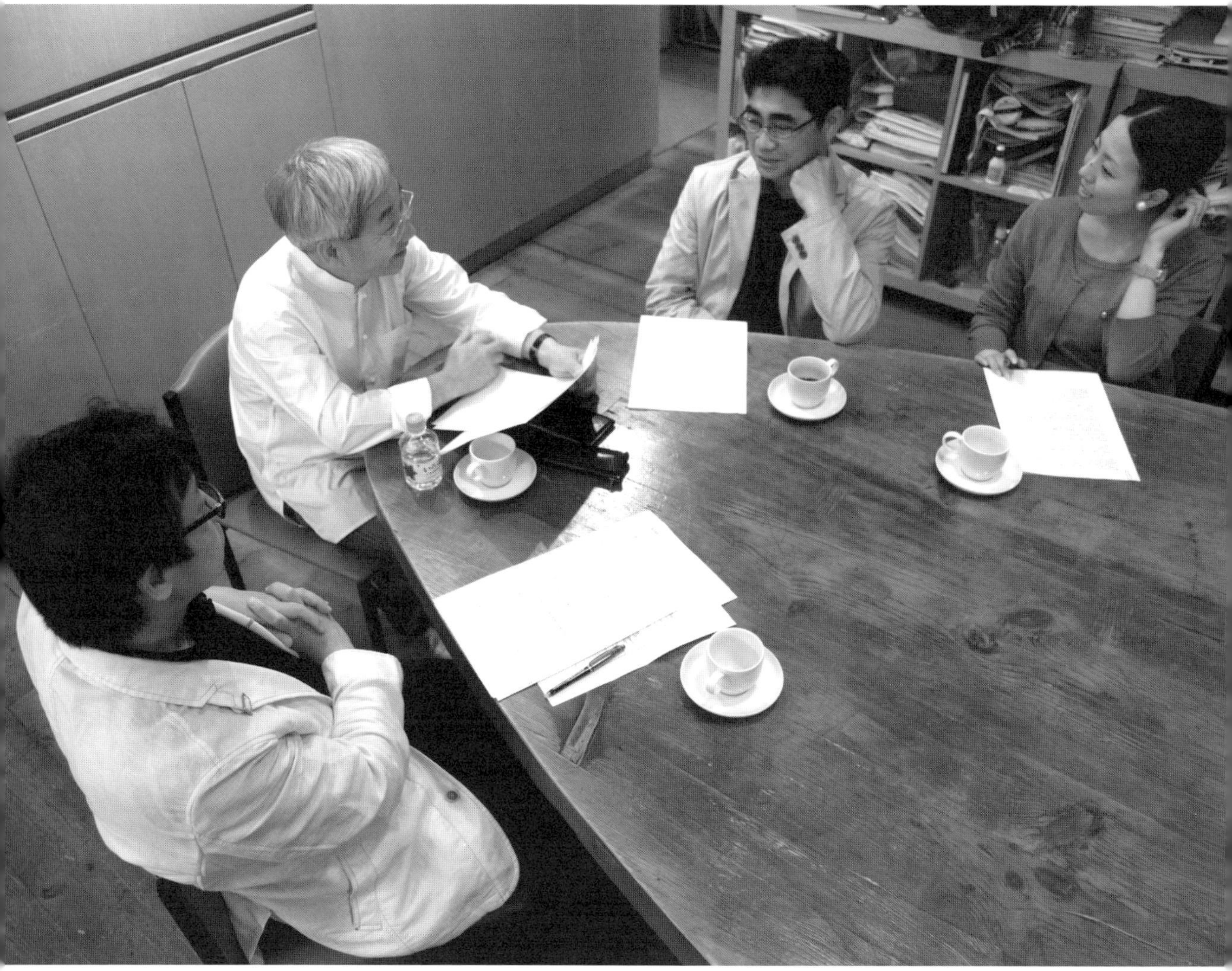

面出　今日は面出ゼミを支えてくれた講師と助手に登場してもらいます。皆さんに光を教えることの難しさや、学生と接点をもつことの楽しさについて自由に語ってもらいたいと思います。まずゼミの運営は、学生と先生の間をうまくコーディネートしてくれる助手という存在があって成り立つわけです。空デの助手というのはとりわけ鍛えられた人が揃っているよね。私は他の大学の事情もよく知っているけれど、空デの助手は一流です。だいたいが情に厚く体育会系で。これだけ鍛えられた助手はなかなかいない。今日は角田さんが助手を代表して参加しているけれど、私のゼミにはこの10年間で3人の助手がいた。初期がお人柄の下田さん、次に厳しくも宴会好きな山下さん、最後がしっかり姉さん役の角田さん（笑）。それぞれに個性的な助手に助けられました。また、私はデザイナーとしてつねに世界中を飛び回っているので、超多忙な私を助けてもらうためにも、私とは個性の異なる非常勤講師の先生にゼミ指導をしてもらっていた。2期生のときにお呼びしたのが林さんで、8期からお呼びしたのが岩井さん。お二人は私にない個性をもっているので、それを期待してゼミ生の異なる筋肉を鍛えてもらいたいと思っていました。

私たちはゼミで何を目指したか

面出　まず最初に、林さん、岩井さんにお聞きしたいのが、光のデザインを教えるというのは、ずいぶん難しい仕事だということです。どうも数学や国語を教えるのとは全く違う。何かとらえどころのないものを相手にして教えている。光はそもそも図面に表しづらいとか、良し悪しも単純に語りきれない。学生との共通言語を探りながら光の本質に迫ろうとしているのですが、明るさの感覚1つにしても「これ明るいだろ」とかいっても素直に納得しないし。その辺のところの教えることの難しさは、他の建築やプロダクトデザインやインテリアデザインとは基本的に違うわけです。むしろ、音だとか匂いだとかと一緒かもしれないような。

林　そうですね、感覚量に関わることですからね。私は2003年のお正月に面出先生からお電話いただ

MENDE　Today, I'd like to talk with the instructors and assistant professors who have long supported the Mende Seminar. I'd like you all to talk freely about the difficulty in teaching light to students and the joy of having opportunities to communicate with students.
In the past ten years, my seminar has had a total of three assistant professors. The first was Mr. Shimoda, a person with a charming personality. Next came Mr. Yamashita, a well-disciplined party lover. And the last was Ms. Tsunoda, a dependable person with a big-sister personality. I have been helped a lot by these assistant professors with distinctive personalities.
We also have two part-time instructors, Mr. Hayashi and Mr. Iwai, who have been teaching our seminar students. They have personalities and abilities I don't have, so I hoped they would train our seminar students in their distinctive ways.

What we aimed for at the seminar

MENDE　I'd like to ask Mr. Hayashi and Mr. Iwai about the difficulty in teaching light design. It's very difficult, fundamentally different from architectural design, product design or interior design. It may be more like sound or smell.
HAYASHI　So, one thing I decided to do was to teach the basics of physics needed to study lighting. Another thing is the sensory aspect. Because we are at an art college, I thought I should provide sensation-related information associated with light and art.
MENDE　When we asked you to help us, Mr. Iwai, we requested you to teach the technical aspect of lighting more clearly, didn't we? I really don't like to teach technical stuff....
IWAI　Because I graduated from a college of architecture, I thought art college students would basically regard lighting as something sensory. So, I thought I was being asked to teach lighting more logically. I tried to teach the basics effectively.

Technology and sensitivity coexist in light education

MENDE　Mr. Hayashi teaches at Tamagawa University, and Mr. Iwai also teaches at the department of architecture at Nihon University's College of Industrial Technology. I also taught at the University of Tokyo's department of architecture, as well as at Musashino Art University, for 10 years. I once gave the same "Light Collection" assignment that I had given to MAU students to University of Tokyo students. Their reaction differed considerably from that of the MAU students.

いて、武蔵美で光を教えるゼミをやっているので、そこで教えてもらえないかということでした。そこで考えたのは、照明を勉強するうえでの物理的な基礎知識をちゃんと教えていこうということと、2つ目は感覚的な部分で、美術大学ですから光とアートという感覚的な部分の情報を見せていこうと思いました。それと私はその当時、照明研究ということを独自に研究課題にしていたので、イメージワードという光を表す言葉を採集していて、小説やコラム、新聞記事などの中で光の表現として使われている言葉にこだわっていました。よく、プレゼンテーションでチッチッチッと光っているとかオノマトペで表される抽象的な言葉もあるでしょう。そういうことを学生たちにも紹介していきました。

面出　なるほど。それは林さんの独特の方法だね。光や影にまつわる言葉や表現は世界中にあるからね。

林　面出先生は、ご自分で体験するいろいろなプロジェクトの中から学生を刺激するだろうから、私は別の切り口からそれを補おうと思って。

面出　私が林さんに講師をお願いできないかと思ったのは、僕が独立する前の会社でずっと一緒だったし、芸大の後輩でもあったし、この人は直感的に面白いと思ったからです。

林　いや、実は予備校でも教えてもらっていたので、かれこれ18歳のときからのお付合いですよね。面出先生はヤマギワの研究所にいたときの上司なのですが、その頃私は主に照明スタンドやシャンデリアのような形態がメインのデザインをしていたのです。その後サラリーマンを辞めるときに、教える仕事をしようと思ったので、光の文化的側面を研究してみたかった。それが武蔵美に呼ばれて少し活きたかもしれませんね。

面出　岩井さんをお呼びしたときには、さらに明確に照明の技術的な側面を指導してもらいたいということでしたよね？　私は技術論を教えるのがあまり好きでないから…。

岩井　私は建築学科の出身なので、最初の照明に入ったのも、テクニカルライティングとかアーキテクチャルライティングといった分野だったから、アート寄りじゃなくてわりとテクニカル派なんですよね。面出先生に武蔵美へ呼ばれたときに、美大の学生

To me, that kind of comparison was very interesting.

IWAI　I'm under the impression that MAU students tend to expose themselves or ardently explore what is interesting to them, expressing themselves sometimes to the extent of making us feel embarrassed.

HAYASHI　Something like an autobiographical novel. For example, there are various units of measure used in connection with light, aren't there?

MENDE　There may be very few art college students who think like a physicist. When teaching light design, it is necessary to logically explain why a particular design is justifiable.

TSUNODA　Is it as a means of teaching in that way that the light box making assignment designed to enable students to experience lighting was given? A light box is a device with various light sources so that they can be lighted differently. Students make their own light boxes by doing wiring and everything by themselves, aren't they?

MENDE　Yes. We often say something like "Learn from natural light." But lighting design can only be learned by using artificial light sources, that is, lamps. I made my students make their own light boxes to enable them to turn on light sources they installed by themselves instead of just looking at such things on a Web page or in a brochure.

Eventually, they have to become confident of their sensitivity to light.

Laborious extracurricular projects

MENDE　I was thinking that we had to step away from what the conventional lighting textbooks say. We just can't create light without visiting the project site. It's essential that we have confidence in our ability to sense light more deeply and delicately than others. What's important, I thought, is the accumulation of your experience in agonizing over what to do with light or the self-confidence resulting from the fact that you have actually seen many kinds of light.

HAYASHI　I really think it's important for a student to actually visit project sites while being a student.

MENDE　They have been given very good opportunities to experience real-world project sites. Every year we had new projects to work on. We really didn't have to go out to find projects. Our seminar was asked to work on new projects one after another.

林 三雄　　Hayashi Mitsuo

1957年山形県生まれ。1981年 東京芸術大学美術学部デザイン科卒業。1884年 東京芸術大学大学院修了。1984年〜1998年LDヤマギワ研究所勤務。玉川大学芸術学部ビジュアル・アーツ学科教授。専門領域｜空間デザイン、照明研究。

Born in Yamagata Prefecture in 1957. Graduated from the Department of Design, Faculty of Fine Arts, Tokyo University of the Arts in 1981. Completed courses in the Graduate School of Fine Arts, Tokyo University of the Arts in 1984. Professor at LD Yamagiwa Institute in 1984 through 1998. Professor at the Department of Visual Arts, College of Arts, Tamagawa University. Specialty: Spatial design and studies on lighting.

というのは基本は感覚的に照明に入ってくるだろうから、それをもっと論理的に教えてほしい、という要望なのかなと思って。基礎をしっかり教えることを心掛けました。

光の教育にはつねに技術と感性が共存する

面出　林さんは玉川大、岩井さんも日大生産の建築や女子美でも教えているし、私も武蔵美と同じ10年間を東大の建築学科でも教えている。武蔵美の学生にだした『ライトコレクション』という課題をそのまま東大生に出したことがあるのですが、反応や成果がだいぶ違うね。私にとってそういう比較論が非常に興味深かった。東大の学生は、武蔵美と比較すると、不要なことまで器用に処理する学習能力がある。特に「巷で光を採集してこい」とかいう課題をだすと、東大の学生は採集するってときに純粋に観察するのではなくて、どうやって見て来たらいいかと、たぶん頭の中で戦略を立てて考えてくるんだよな。それはそれで面白い部分もあるけれど、私はその方法を叱責することが多い。「そういうことじゃないだろう」って。要するに「素直に自分の感覚機能を広げろ」「頭の中で考え過ぎるな」というわけで。その点で武蔵美の学生はひたすら純粋だね。私のいうことに感覚的に反応する。

岩井　なんか武蔵美生のほうが自分をさらけ出してくるというか、自分の興味のあるポイントを追求していくというか。こっちが恥ずかしくなるくらいさらけ出した表現をするよね。

林　わりと私小説に近い感じの表現とかね。たとえば、光で使ういろいろな数字を伴う単位ありますよね。ルクスとかカンデラとか、そういうのは説明しても当然わからないしね。彼らが感覚量を数値化することに興味を示さないことがずいぶんありました。

岩井　明るさには輝度と照度との2つあるじゃない、

TSUNODA　　Unlike ordinary classes at the university, Mende Seminar's classes are actually projects for specific purposes. It's true it was difficult because each class called for project-specific preparation and approach, but each class gave me a new surprise and a moving experience. I was involved in those projects as if I were a project team member, so I feel I was able to learn together with the Mende Seminar members.

IWAI　　I also got confused at first. I had been told that all I had to do was to conduct classes, but when I came here, I was surprised to find that there was so much fieldwork. But I think we were lucky we had so many projects to work on.

TSUNODA　　That's true. There are many real-world projects. There are project sites and clients, and we have budget restraints. Not many students have opportunities to have such experience under the guidance and with the assistance of teachers. It's only at the Mende Seminar that you can do things like that. So, it's amazing how students grow up every time a project is completed, sometimes to the extent of amusing.

But I think that students had hard time. Things occur at a speed they have never experienced, and they are not allowed to stop or give up. Prof. Mende is always there, meticulously checking on what they are doing. I often saw students who actually looked pale.

MENDE　　Yes. I don't want to scold students without giving them a chance to explain, but I also feel that I should tell them the truth. Students at first are not very tough. So, a part of me thinks that I should be tolerant and should not be too demanding. It's difficult to design light, isn't it? Saying that I don't like it is not enough. It's necessary to explain objectively why a particular design is not good. There were times, though, when I wondered if I was being too permissive. I thought a more sound approach was to say that something is no good if it's intuitively unacceptable. I try to scold students affectionately without deeply hurting them.

Both students and teachers shed tears and laughed a lot

MENDE　　Ms. Tsunoda was like an intermediary between us, professors and instructors, and students. What gave you the hardest time? I mean, you shed tears a lot. Maybe tears of joy sometimes?

そこに関わっているものが輝度だったら反射率が関わっているとか、そういう単位だけで説明するのではなくて、どういうことなのかということを紐解きながら、でも物理的に教える必要があるんだよね。

面出　美大には物理的な頭をもった学生は少ないかもしれないね。だから私は学生にいつもいうんだけれど、デザインはいつも科学と芸術の間を行き来するわけだから、「自分は感覚派なんだ」とかバカなことはいわずに、テクニカルなことも理屈や法則もある程度説明できるように学んでおけと。光のデザインを教えるというときには、デザインの正当性を理屈の中で説明できる必要がある。

林　単純に光は足し算なんだよと教えます。暗かったらもう1つ足せば、足しただけの明るさが取れるとか、距離的に2倍離れると光の量が4分の1になるとかね。そういう感覚的なことを越えた物理的な抑えというのも大事かなと思いますね。

面出　だから二人の講師の先生を招いたことで、ある意味で彼らの苦手としている分野のことも、工夫しながらぐっとねじ込んでもらうことができたことが良かったと思っています。

角田　その教える工夫の中から生まれたのが、照明を体感しながら教えるライトボックスの制作ですか？　たくさんの種類の光源を取りつけて点灯できるようにした装置ですが、学生が自分で配線したりしてつくったのですよね？

林　そうだね。あれは助手が下田さんの頃だったかな。面出先生と「教えるにも、まず光源がないと話にならないよね」という話になって、どれだけ種類があるのか、まずは本物で見せようということでつくったのですよね。

面出　そう。「自然光に学べ」とかいっているけれど、やはり照明のデザインというのは人工的な光源＝ランプがあって学習できる。それもWebとかカタログの中で見るのではなく、自分で取りつけた光源を点灯させてみよう、ということでライトボックスをつくらせた。光を学ぶことは、自分の感じられる能力ということに、最終的には自信をもってもらわなければいけないわけだよ。自分の感覚機能を鍛えるためにいろいろなトレーニングを学生にもしてもらうわけだけれど、その中で生身の光を体験することで、そ

TSUNODA　Oh, yes. Happy tears in most cases. The presentation of light is practically instantaneous, but I was always in the position to know how much time the students had spent in dealing with light and people. In almost all cases, I burst into tears when a project came to completion.

MENDE　I seldom shed tears, but there were times when I found myself close to tears, in some cases because I was happy and in others because my heartfelt desire was not fulfilled. I felt awkward because the students saw me in tears. To tell you the truth, there were many times, other than those I've just mentioned, when I thought my approach might have failed.

Light must not be virtual

MENDE　Today, students frequently use computers and mobile devices. Many students work on computers rather than talk with teachers. That often makes it difficult for us to teach light design.

HAYASHI　Yes, indeed. The Internet is sometimes harmful. The Internet may be useful when you can find something unexpectedly so as to inspire you in one way or another, but it can also be a factor that makes it difficult for you to write down your own thoughts in your reports.

MENDE　I don't want to blame students, but we live in a very convenient society in which we can obtain various information even while walking. You can view anything you want, even an aurora occurring on the other side of the earth, on this very convenient device. Actually you are not seeing it, but you think you saw it or you know it ….

IWAI　Prof. Mende conducts classes designed to show real things at real project sites to students, and I think that's the most important part. In short, the idea is to enable students to experience the difference between information and the real thing by taking them out of the university and showing the real thing. I think that makes a big difference.

MENDE　Absolutely. It must not be virtual. Reading books and looking at beautiful pictures is not enough. Looking at the glittering computer screen is not enough.

HAYASHI　That's the important part in designing light. I believe you can't acquire the sensitivity to make intuitive design decisions just by looking at things on the Internet. I want students to challenge the established values.

MENDE　That's right. Even now, I sometimes find myself questioning my own long-established approach to lighting design. I fear the so-called common sense. To design light is like removing the scales from your eyes every day, isn't it?

れぞれの感じ方の差異を知ることになる。いろいろなところに感じ方の違う人がいる。その差異を尊重するのは光のデザインの中で大切なこと。それでないとね、西洋合理主義のような1つの文化の優越性を押しつけたがる人もいるからさ。

体力勝負の課外プロジェクト

面出　この本の中でも僕は学生に光を教えるときの7つのキーワードを挙げていて、武蔵美に来たときにずっと思っていたのが、これまでの照明の教科書に書いてあることから離れないといけない、と思っていたのですよ。私も照明の教科書をたくさん読んだけれど、机上で学習できるデザインの深さは知れているから、どうしても現場で光と格闘するということになるんだよね。現場に入ることを抜きに光を創作することはできない。私たちはほかの人以上に、自分が光を深く繊細に感じられる人間だという自負が不可欠になる。光に触わって苦悩した回数、またはたくさんの種類の光を目に焼きつけているという自負、そこが大切なのではないかと思ったの。だから面出ゼミでは、そういう体験の積み重ねを大事にしていたと思う。その結果、ゼミ生は個人の作品制作みたいなものをあまりしていないのではないかな。

林　わりとそうですね。共同のプロジェクト多いですからね。個人作品をつくる暇もない。

面出　一般的には照明課題というと、照明器具をデザインしなさい、とかがわかりやすいのですが、うちのゼミだと勝手にデザインさせる課題がなかなか与えられない。安易なデザインするより生身の光と格闘することが多いよね。

林　学生のうちに実際の現場を与えられるというのは、すごく大事かと思うのですね。ライトコレクションの課題では、それなりに時間をかけてじっくり光を見る。光を観察し体験する訓練を積んでから、その後でいろいろな課外プロジェクトに駆り出される。明治神宮のライトアップとか、小池先生や新正先生の実際の展示空間をスポットライティングしたり、岩室温泉郷などではあかりの町おこしまで現場で提案する。現場しかないかなという気がしますね、光を勉強するというのは。

I wonder why students these days are so gentle.
How to bring out the individuality of students

MENDE　There's something that makes me wonder. Students who happened to gather at the Mende Seminar are often said to be unique and gutsy. I wonder why they have such characteristics.

HAYASHI　That's because you are good, professor. You have the ability to bring out the individuality of each student by saying something like "You're weird." Talking with students about something not related to classes, you bring out unique characteristics in the students.

IWAI　Kind of bringing out the best in each student without paying much attention to weaknesses. Basically, Prof. Mende is good at praising students, tactfully bringing out the best in each student. It's like making each student's strengths visible to others.

TSUNODA　Spoken to in that way by Prof. Mende, each student finds his or her place in the seminar group, and that's where they can start working on whatever they want.

HAYASHI　It may be as simple as saying something like "I'm counting on you for party planning." when he hears that a particular student likes drinking. Something as simple as that enables each student to find her or his place.

HAYASHI　In autumn every year, just before a new seminar season for third-year students begins after the summer vacation, there's an annual seminar camp in which third- and fourth-year students participate. A camp of two or three nights in a place like Kiyosato or Oshima. I feel that Prof. Mende effectively uses that occasion to win the hearts of the students.

MENDE　That's a long-established annual event. One of the traditions of the Mende Seminar. To make a small memory. It's important to share the food out of the same pan as an old saying says. One characteristic of our seminar is that third- and fourth-year students mingle together to do things jointly. The summer camp may indeed be a contributing factor as Mr. Hayashi said.

TSUNODA　At first, Prof. Mende stirs up classes and students vigorously as when making whipped cream. And in time you find a much larger volume of much more beautiful whipped mixture. I was feeling that way while participating in the Mende Seminar and found myself deeply involved without knowing it. Something like that.

HAYASHI　Exactly. Eventually, Prof. Mende makes a final decision, but the students don't think that the professor made a decision for them. Instead, they are satisfied with the decision. That's an interesting part.

MENDE　That's a difficult part. The results of semi-

面出　非常にリアルな場所が与えられてね。よく毎年いろいろなプロジェクトが矢継ぎ早にあったよね。こちらから探してお願いするわけでもないけれど、次々に新しいプロジェクトがゼミに依頼されてきた。

林　さっきライトボックスの話が出ましたけど、やはり光源だけ見ているだけでは、演色性とか色温度とかグレアの差などはわかりにくい。現場に入って初めてそれが如実にわかるのですね。実際の空間で光を操作していくという体験ができたというのは、毎年毎年良かったと思いますね。

面出　私たちは学生を頻繁に現場に連れて行くことを由としていたけれど、角田さんから見てどうだった、大変だったのではないですか。学生はつねに俺には弱音を吐かないけれども、もしかしたら学生も角田さんも「大変なことになった…」と愚痴っていたのでは？

角田　面出ゼミの授業は、大学で行っている通常の授業とは違って、1つ1つプロジェクトになっているんですよね。だから、もちろん毎回準備も対応も同じわけにもいかないので大変だったところもありました。ただ、それより毎回驚きや感動の連続で、私もメンバーの一員みたいな気分で関わってこられたので、面出ゼミと共に私も成長してこられたような気がしています。

岩井　僕だって最初、面食らいましたよ。普通に授業だけしてくれればいいと聞いて来てみたら、江戸東京たてもの園のプロジェクトとかシンガポールのキャンドルナイトとかが始まっていて。現場作業とかまであって「これって全然、授業の範囲越えてるじゃん！」って（笑）。結局、初めに面出先生と立てたスケジュールなんて全くその通りなんかにはいかなくて、仕方なく4年生になってから追加授業したりしてね。でもなんだかんだいっても、あれだけ多くのプロジェクトがあるということはすごくいいことだと思います。

面出　どうしてなのか自分でもよくわからないのだけれど、毎年そういうのがあって。学外に連れ出したり、責任ある仕事をみんなでやろうということになったんだよね。

角田　そうですね。ほんとう、リアルなことが多いですよね。現場があって、クライアントがいて、予算が

nar activities are not my own achievements, and there are times when in a sense I allow students to make a big mistake instead of achieving a big success. Actually, there were many times when I said, "Things didn't turn out as you expected, did they? Frustrating, isn't it?" But, finally, I can't compromise when it comes to making a decision as to which way to go. That's why I believe that in an educational environment, teachers should try to learn from students.

Genes of Mende Seminar inherited?

MENDE　The reason why I've decided to record the ten years of this seminar is that I want to pass what I tried to teach about light design on to next generations. I'm sure many future students will learn about light design. So, I'll be happy if what we can pass on—something that can be described as "the genes of the Mende Seminar"—helps future students and teachers in any way.

HAYASHI　As Prof. Mende said in advocating the renewal of In Praise of Shadows, which Jun'ichiro Tanizaki, grieving over the modernization of lighting environments, wrote early in the Showa period, I believe the time has come when we should renew lighting environments. Maybe we should warn the world about the current situation.

MENDE　Instead of just reading the classic In Praise of Shadows and looking back on the good old days, we must criticize the modern age now just as Tanizaki did in the past. I wanted to help students learn to look critically at the world. Maybe we should enjoy low-tech lifestyles while learning state-of-the-art technologies …

MENDE　To me, in a sense, the seminar activities in the past ten years have been like child-rearing. I did whatever I could think of when I was unable to go well with students. I was concerned about the possibility of their becoming delinquent, but art college students should be somewhat delinquent or avant-garde. I was always trying to see things in the same way that students do.

First of all, I forced them to make effort. Then, I tried to make them talk in an unabashed and arrogant way. To do that, I tried to make them criticize themselves, friends and society in general.

The most important thing, however, was to have good times, overcome difficulties by drawing on the memories of the good times, and cooperate with others in a comfortable way instead of working alone. These may be the genes of the Mende Seminar.

岩井達弥

Tatsuya Iwai

1955年東京都生まれ。1980年日本大学理工学部建築学科卒業。1980年TLヤマギワ研究所入社。1996年岩井達弥光景デザイン設立。国際照明デザイナーズ協会会員。日大、女子美、武蔵美非常勤講師。

Born in Tokyo in 1955. Graduated from the Department of Architecture, College of Science and Technology, Nihon University and joined TL Yamagiwa Institute in 1980. Founded Iwai Lumimedia Design in 1996. Member of the International Association of Lighting Designers. Part-time instructor, Nihon University, Joshibi University of Art and Design and Musashino Art University.

あってというように、なかなか学生時代にそんな経験できないことを、先生方のアドバイスやフォローのある状況で行うことができるというのは、面出ゼミならではだと思います。だからプロジェクトが1つ終わるごとに、人間的にも成長して大人になっていく様子に驚きます。見ていてとても面白いくらいに。

岩井 普通の学校ではできないよね。だからそういう意味で、面出ゼミの学生は幸せだったと思う。だから、ほんとうに意欲のある子は伸びるよね。

林 学生たちは相当大変だったんじゃない？やはり責任感の有る無しとかね。

角田 学生はとにかく大変だったと思います。今まで自分たちが体験したことのないスピードで物事が進んでいて、止まることも降りることもできない状況の中で、面出先生の厳しいチェックが稲妻のように入って、青い顔しているときとかしょっちゅうでしたよね。

面出 そう、それで私は頭ごなしに叱らないつもりなんだけれど、ほんとうのことをいうべきだとはどこかで思っていて。初期的には学生というのはそれほどタフじゃないから、どうにか温かく、少し手加減しなきゃいけないとどこかで思っている。光のデザインって難しいじゃないですか。自分の好みの問題だけでなく、客観的になぜこれがダメなのかを明確に説明しなければならない。だけど、あまり学生のために声色を整えていてばかりでもいけないかな、と思った瞬間があって。やはり、直感的にダメなものはダメというほうが健全だと思ったんだよね。決定的に傷つけないように、気遣いながら愛情をもって叱るわけです。

角田 毎回グループでの制作となると、リーダーになる学生はプロジェクトの内容だけではなくて、人間関係とかでの難しさとかもつねに抱えていて、先生には制作のことの相談で一杯一杯で、泣き言なんかいえなかったぶん、私たち助手にはそういった面の相談もつきものだったような気がします。とにかく学生の話を「うんうん」とひたすら聞くことで、自分で解決して立ち直っていくときもあれば、可愛いがゆえに厳しくもしたことも多かったように思います。だから学生とは仲も良かったと思っていますが、結構私も怖がられていたみたいですね（笑）。

面出 そうだよね。私が40年前くらいに学生やっていた頃と、今の学生の雰囲気はもちろん違うだろうしね。そこのところは、今の学生は恵まれていると思うこともあるけれど、飽食の時代に生まれたことの難しさもある。僕らはもっと単純だったからさ。だから今の学生もずいぶん大変だと思うよ。

学生も先生もよく泣き、よく笑った

面出 角田さんは、私たち教授や講師陣と学生との真ん中に入っていて一番苦労したことって何？あなたは結構泣いたじゃない（笑）。

角田 面出ゼミには結構泣かされましたね（笑）。いろいろな泣きがありますけれどね。

面出 嬉し泣きもあった？

角田 はい！感動泣きのほうが多いですね。みんなに見せる光はパッと一瞬のことでも、そのときのために学生がどれくらい時間をかけて光や人と向き合っていたか、というのをいつも陰から見ていたので、どのプロジェクトでもいつも本番ではたいてい号泣していました（笑）。まあ、感動泣き以外には、どんなにこちらが学生に向き合おうとしても、こうもうまくいかないものかと、もどかしくて、悔しくて…、というときもありました。

面出 私はめったに泣かないけれど、すごく嬉しいときだけでなく、自分の深い思いが叶わなくて悔しいときにも目頭を熱くしたことがあったよね。卒業制作の最終講評のときにさ、当人は頑張っていたんだけれど講評時になってもでき上がらないんだよ。

角田真祐子

武蔵野美術大学空間演出デザイン学科研究室助手/1985年神奈川県三浦市生まれ。2008年武蔵野美術大学空間演出デザイン学科卒業。2008年武蔵野美術大学空間演出デザイン学科研究室教務補助着任。2009年同研究室助手就任。同年デザインユニットminna設立。

Mayuko Tsunoda

Assistant Professor, Department of Scenography, Display & Fashion Design, Musashino Art University.
Born in the city of Miura, Kanagawa Prefecture in 1985. Graduated from the Department of Scenography, Display & Fashion Design Musashino Art University in 2008. Was assigned a job of supporting instructor of the Department of Scenography, Display & Fashion Design in 2008 and was appointed an assistant professor of the department in 2009. Founded minna, a design unit, in 2009.

頑張っていたから、きっちり終わらせて褒めてあげたいと思っていたんだけどね。これはね、やはり指導の仕方が間違ったかなと反省した。学生に涙見られちゃって不味いなと思ったよ。実はこれ以外にも、「私のやり方が不味かったのかな」とか思うことはたくさんあった。

林　まあ、教えているとそれはよくよく思いますけれどね。毎年違うプロジェクトを与えるので、去年こうやったから今年はこうしようとか、できることばかりではないのがきついですよね。

面出　まあ、でも今、角田さんがいったみたいなことの1つ1つができていくというか、当たり前に苦労を重ねてさ、辛ければ辛いほどそれがどうにかことを迎えたときの安堵の念というか、またはやり終わった後、傷だらけなんだけれども、「おう、俺たち少し筋肉ついたぞ！」みたいな実感があると、やはりやって良かったなという、やらなくて逃げたのではもっと辛かったなということだけなんだよな。

光はバーチャルではいけない

面出　学生が頻繁にパソコンやモバイルを使うようになってからは、先生と話すのではなく、パソコンに向かっている学生が多かったりしてさ。私もびっくりしたのだけれど、「本音をきちんと話してくれよ」と叱ったら、「先生、私のブログ読んで下さい」というんだよ。要するに対面してなかなか本音をいえない。コンピューターや携帯、ブログばかり相手にしていると、目を見てお互いの体温を感じて話をするという、この距離感をものすごく苦手にしている学生がいる。全員がそうではないけれど押しなべて多いね。

岩井　確かにそういう人は多くなってきている。私もときどき困惑しますよ。

面出　だからそれは光のデザインを教えるときに困ることが多い。「たくさんの光を観て来いよ」と注文すると、次のときにすごくきれいな夕暮れ時の資料を揃えてきてさ。「おお、すごいな。これはどこで撮ったんだ？」と聞くと、「webでちょっと拾ってきました」というんだよ。びっくりしちゃって破り捨てようかと思った（笑）。「光のデザインをwebの中で探すな！」って叱ったさ。

林　そうなんですよね。インターネットが有害な部分をもっていることもあるし、逆に自分の予期しないものが検索するとパッと出てきて、それはそれでインスピレーションになっていい部分もあるのだけれど、自分の考えがレポートとして出てこない要因でもあると思うんです。

面出　学生を攻めるつもりはないんだけれども、ある非常に便利な社会が訪れてきて、いろいろな情報が歩いていても即座に手に入るみたいなことになった。地球の反対側で起こっているオーロラ現象でも何でも、この便利な機器の中で見ることができる。ほんとうは見ていないけれど、それを見た、または知ったということになるわけで…。

岩井　だからそれはさ、情報として得るのはいいけれど、結局そこからその情報を得て本物を見たいという欲求や発想が出てこない。ほとんどの学生が画面で見ただけで満足してしまうことがある。だから、面出先生は本物を現場で見せるための授業をやっているから、多分そこに最大の意味があるのではないかと私は思う。要するに、学内だけで終わらせるのではなくて、外に連れ出して見せることで、情報と本物の差というのを体感する。そこにこそ、すごい意味があると思うのです。

面出　絶対そうだと思うけれどね。やはりバーチャルじゃいけないし、本ばかり参考にきれいな絵ばかり見ていてもいけないし、光る画面ばかり覗いていてもいけない。今は社会的にはとても平和なので、学生同士の争いごとがうんと少ない。それも問題なんだ。私の青春時代には、自分と友達との差異はどこに

あるか、ということだけが非常に大切だったからね。

林 そこが光をデザインするうえで大切なところで、デザインを直感的に決めるときの感覚というのは、インターネットで見ているだけでは身につかないんだろうな。既成の価値観をグラグラグラっと揺すってほしいですね。

面出 そうなんだよ。デッサンだって、描いている途中でときどき逆さまにして見たりするじゃない。学生には裏返しにして作品を見るとか、逆立ちして世の中を見てみるとかさ、いろいろな所作を試してほしいわけ。私なんか今でもそうだけれどもね、日常の中で染みついた照明デザインを疑ってかかる自分がいる。常識というのが恐ろしい。光のデザインで、毎日が目の鱗落としみたいなものじゃないですか。

岩井 今の社会というのは全部、便利に組み立てられているから、いろいろなことがすぐにできてしまう。その便利さ、安易さも手伝っているのかもしれませんね。

面出 なぜ、今の学生というのはあんなに優しいんだろうね。

角田 何でですかね。面出ゼミ生も、特に男子は優しい子が多い傾向にある気がしますけれど、優しい中にも個性が立っているほうだとは感じますね。私がちょっと不思議に思っていることは、個人個人だといい意味で変わり者が多いのに、みんなでプロジェクトとして動くときにはよくまとまるなあと感心します。自己主張が強いだけではないというか、協調性はゼミを通じて育っていっているのかもしれないですけれどね。普段の様子からすると、考えられないくらいの学生とかいますね。

学生の個性の引き出し方

面出 でもどうなんだろうね。毎年のゼミ生の違いはあるけれども、面出ゼミに偶然集まった学生たちは一人ひとり個性があってガッツがあるといわれるよね。そういう特徴があるのは何でなんだろうね。

林 それは面出先生がうまいんですよ。一人ひとりに「この子、変な奴だな」とかいって、いろいろ引っ張り出す力があるんですよ。僕は面出先生の授業を手伝うようになって「あっ、うまいなあ」とすごく参考にさせてもらっているんです。わりと学生と授業に関係ない話とかしながら、個性を引き出していますよね。3回くらい授業をやっていくと、だいたい「この子はこんな感じで、こんな個性があるな」とかわかってきますよね。

岩井 あまり短所は見ずに、長所だけ伸ばしていくという感じですよね。

林 たとえば「この学生はコンピューター関係とかプログラミングとか得意そうだな」とかがわかったら「お前それ得意だろ！どんどんやってみて！」といった感じで。

岩井 面出先生は基本的に褒め上手でしょう。学生の得意技とか長所を引っ張りだしていく。みんなにそれぞれの長所を見えるようにしてくる、という感じですかね。

角田 そうやって面出先生に声を掛けられた学生は、ゼミという集団の中で自分の居場所というか、ポジションを見つけて、そこを起点としてより頑張ることができているという感じはしますね。

岩井 能力の高いところをピックアップしようとしている感じかもね。

林 単純にお酒が好きだって聞いたら「飲み会の企画はよろしくな」みたいな感じとか。そんな大したことないことだけれど、それで学生には居場所ができているんじゃないかな。

岩井 できないところを「やれ、もっとやれ」といっても嫌になっちゃうし、もう大学くらいまできたら、長所を伸ばしていくやり方のほうが僕はいいと思いますけれど。

林 秋から3年の新しいゼミがスタートする前、ちょうど夏休みの終わりに3年と4年合同のゼミ合宿があるじゃないですか、清里だとか大島で2～3日寝泊まりして。あれで結構面出先生は、学生たちをグッとつかんできているのだろうなと感じますね。

面出 あれは結構初期から行っているからね。面出ゼミの伝統の1つになっている。小さな想い出づくりさ。みんなで同じ釜の飯を食う体験は大切だね。うちのゼミの特徴は3年と4年が団子みたいになりながら、先輩後輩って当たり前に協力しながらやるという体制ができている。それは林さんがいうように、夏の合同合宿のお陰があるかもしれないね。角田

さんは面出ゼミの助手でどうだった？

角田 たまたま担当助手三人とも、面出ゼミの卒業生ではなかったのですが、私は面出ゼミの一員だったかのような錯覚を起こすくらい、深く関わってこられたとは感じています。

面出 ファミリーの中の一員として、助手にもちゃんと役割を与えてね。知らぬ間に助手まで僕の罠にはめていたわけだ（笑）。

角田 面出先生って、授業とか学生のことを生クリームをつくるときみたいに、初めは一生懸命かき混ぜるんですよね。でも、気づいたら泡立って何倍もの量や美味しさになってまとまったみたいな。私はそんなふうに感じつつ面出ゼミと関わっていくうちに「気づいたら私まで巻き込まれていた！」みたいな感じです。

林 そうなんだよね。結構最終的なところで面出先生が「よし！こうする！」というようにやっちゃうんだけれど、学生はそれを先生に決められちゃったというのではなしに、納得ずくで「そうだよね！」とまとまっちゃうというか。そこのところが面白いところかもね。

面出 難しいところなんだよ。ゼミの成果というのは私個人の成果ではないし、学生にある意味で大成功ではなくて大失敗させることもある。「思うようにいかなかっただろう、悔しいよな」ということもたくさんあったわけですよ。それはある範囲の中で教育的にも大切なことなので、全部が自分で決めちゃおうとはしないし、イライラしながらでも勝手にやらせるわけです。だけど最後になって「やはりこっち側にいくべきだ！」、みたいに私が思うところをいい加減にはできない。だから教育の現場は、教える側が学ぶ側から教わることを由とすべきなんだ。そこのところがなかなか難しいね。生クリームの撹拌される中にいながら必死に何かをやっていって、面出ゼミを卒業した学生は破格に逞しくなってほしいといつも思っているからね。

林 毎年初めてのゼミのときに、面出先生がいう、面出ゼミの鉄則みたいなのあったじゃないですか。「時間を守る、挨拶をする、清掃をする」でしたよね。初めて聞いたとき「え、そんなことというの」と驚いたんですけれど、結局社会に出て、その3つが一番大事だということですよね。

岩井 それと面出先生は出欠をとるとき、名前だけを読み上げずに一人ずつ会話しながらとるじゃない。あれもいいよね、あれがコミュニケーションという感じで。なかなかうまく真似できない（笑）。そういうときと、かき混ぜるときと両方なのがいいんでしょうね。

　　　　面出ゼミの遺伝子は残ったか…

面出 うちのゼミの10年間を記録にまとめるのは、光のデザインを教えようとしたことを次の世代にも伝えたいと思ってのこと。これから先にも光のデザインを学ぶ学生はたくさんいるはずだから、その学生たち、先生たちに対して面出ゼミの遺伝子みたいなものが、少しでも役に立ってくれたら嬉しいわけです。これまでに話したように社会がどんどんバーチャル化して、エネルギー不足に見舞われて、孤独な人が蔓延してきて…と、様々な閉塞感を伴う現象が起きる。その中にあって、光のデザインというのは大きな期待を抱かせるものなのです。光だけでなく、私たちのゼミでは闇や陰影の価値を語り続けてきた。だから「面出ゼミは暗いんだ」と冗談交じりにいったりもする。「光は影だ」とか「明るさは暗さだ」とか、「光は移ろいの中に」みたいな特徴的な概念をたくさん語ってきた。それらが全部、ゼミのDNAとなってきたんだろうね。

林 先日のセミナーで面出先生がお話されていたことで「LEDとコンピューターが出てきて、昔より手軽に看板屋さんでもグラフィックデザイナーでも光を扱うことが簡単にできてしまうから、すごく下品な街の景色になっちゃう」ということを話されていて、全くそれに同感です。そこのところを光環境リテラシーという視点で考えて、光のデリバリーの感度を洗練していかないといけないだろうなと思いましたね。

面出 照明を文明論でなく、もっと深く文化的に考える。やはり光に向かう真摯な態度が必要だね。照明デザインの神髄を極めるという姿勢がないと、みんなが簡単にLEDやコンピューターの中だけで照明デザインを扱ったりするからね。

林　面出先生がおっしゃっていた「陰影礼賛を刷新せよ」というように、昭和の始め頃、谷崎潤一郎が灯り環境の近代化に嘆いて書いたわけですけれど、またそのときのようにリニューアルするべき時代がきているのだろうと思います。世の中に警告したほうがいいのかもしれませんね。

面出　『陰影礼賛』という名著があり、それを読んで昔は良かったと回顧するのではなくて、その谷崎が嘆いたその時代と同様なことを、私たちが今ここで現代を批評しなければならないのです。その批評するという目を学生たちにももたせたかった。それでないと技術のみがどんどん発展していって、それはもう否定しようがないし、それが悪いわけではないから、コンピューターに使われるのではなく、我々がコンピューターを使いこなしていくためには、陰影礼讃を現代に刷新する必要があるのです。ハイテク技術を学びながら、ローテクな暮らし向きを楽しむべきでは…。

岩井　ローテクというか、もっと人に近いということだろうね。技術が人に近づかなければ、と思いますね。

面出　ゼミの10年間は、私にとっての子育てだったと思っているところもあるんですよ。私が武蔵美に教えにきたのが52歳のときで、ちょうど大学生の子供をもつ親の歳くらいだった。そこから10年で今62歳。そろそろ子育ても終わりかなという思いもあって、今年で退任することを決めたわけです。10年間でたくさんの学生と関わってきて、子育てというのはほんとうは私自身の成長剤だったように思います。学生とうまくいかないようなときには、あの手この手といろいろ考えて手を打った。学生が非行に走らないように心配もするけれど、多少の非行やアバンギャルドがなければ美術大学らしくない。私も学生と同じ目線になろうとつねに思っていた。

角田　私個人的には、まだまだゼミを続けてもらいたいと思っていたところもあったのですけれど、以前に「子育て」のことを面出先生から聞いたとき「ああ、なるほど」とすごく納得してしまって。今までのゼミ生と面出先生との日々が走馬灯のように頭の中を流れていったのを覚えています。

面出　毎年毎年が思い出深くて、よくもこう様々な、個性の強い学生がいたと思うよ。私も親の躾のつもりで、細かいこと、喧しいことをいい続けてきたけれど、学生もよくついてきてくれた。私は彼らにまずは努力することを強要したね。その上で、できるだけ恥ずかしがらずに偉そうなことをいうように、そのためには自分や友人や社会を批評させようとした。しかし何よりも大切なことは楽しく時間を過ごすこと、辛いときにもめげずにそれを糧としてしまうこと、自分一人だけでなく他者と気持ちよく連携すること、それが面出ゼミの遺伝子かな。光のデザインを学ぶという所作は、その行為自体が抽象的であったり不可思議であったりするので、取り立てた職能教育にはならなかった。それがかえって良かったのかも…。人のためにデザインをするという凛とした姿勢を保ってほしいと思うね。

2012年6月10日
武蔵野美術大学　空間演出デザイン学科研究室にて

2012年9月1日、OBを交えた最後の清里合宿
Last annual camp in Kiyosato with alumni, September 1st in 2012

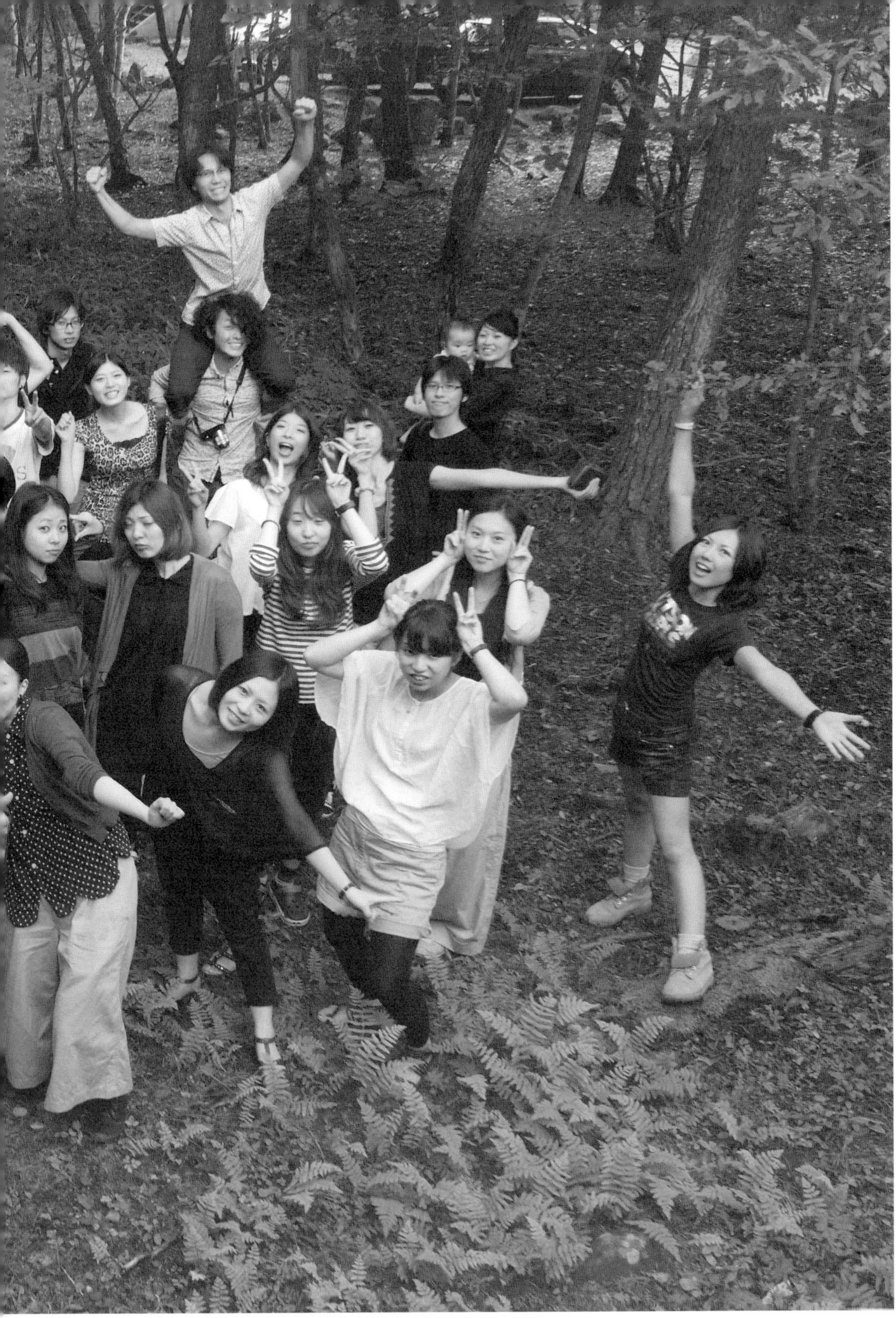

あとがき
Afterword

10年間の子育ては終わった。ここ数年の胸の内を正直に語るとそんな思いがする。

　私はいつも誰かのためになる仕事を心掛けてきた。それはデザイナーという職能に付きまとう習性である以前に、私の生きかたそのものが常に私以外の誰かのために向けられていたのかも知れない。たくさんの仲間と志を共有するのが好きなのだろう。徒党を組むことを得意としてきた。武蔵美に赴任した2002年の4月からの10年間、私は目の前を通過する多くのゼミ生と交わり、喜怒哀楽を共にし、そのことは私に多くの知恵と新しい思想をも与えてくれた。しかし楽しい子育てに区切りをつけようと決心した。

　退任するきっかけはデザイン教育をもっと深めたいと思ったからでもある。世界中を駆け巡り多忙を極める私の日常には真正面からデザイン教育を考える暇もなかった。退任を決めたと同時にこの10年間を顧みる仕事、すなわちこの本『光のゼミナール』の企画を開始した。この出版を通じて私なりのデザイン教育論を総括してみたかったのである。また、学生たちにとっては面出薫は何者であったのかも知りたいところだった。

　1年前から取り組んだ学生たちとの最後の仕事がこの本の出版である。楽しい子育てを自ら終了するには多少の勇気を必要としたが、このような出版の機会が与えられたことに心より感謝している。

　私は最後までたくさんの人々に囲まれて良い仕事をさせてもらった。この本の出版も面出ゼミOBによる編集委員会を作り月1度のワークショップを重ねた成果である。本業を持ちながら多くの時間を割いて参加してくれたOBたち、とりわけ編集委員長の加藤直子さん、デザイン担当の中村将大さん、ゼミデータ担当の角田真佑子さんには一方ならぬ協力をいただいた。また私の良き先輩である杉本貴志さん、小池一子さん、同僚の小竹信節さん、原研哉さん、佐藤卓さんには暖かい寄稿文をいただき感謝に堪えない。更に我儘な私たちを厳しく暖かく指導していただいた鹿島出版会の相川幸二さん、デザイン協力の高木達樹さんにもこの場を借りて深くお礼を申し上げる。なお、本書の刊行にあたっては、武蔵野美術大学より出版助成をいただいたことも付記させていただく。

2013年1月6日

面出　薫

The ten-year child-rearing was over. This is the honest feeling I have carried for these years. I have always made sure to work for somebody else. I might say that such a habit is inherent in the profession of designer. Rather, my way of life may have always been directed toward someone else. I may like sharing interest with a lot of associates. I was good at forming teams. I met and shared feelings with numerous seminar member students who came and went in the ten-year period after I took a post in Musashino Art University. The experience gave me a lot of wisdom and new ideas. I decided to draw a line in pleasant child-rearing and resign in March 2013.

　What motivated me to resign was a desire to reinforce design education. I had no time to squarely think about design education while I was hustling and bustling all over the world. When I made a decision on my resignation, I started planning on "Light seminar", the book in which I would recollect the period of ten years since 2002. I wanted to summarize a theory of my own on design education by publishing the book. I also hoped to know who Kaoru Mende was to students.

I have worked on the publication of the book with students for one year as the last job. Discontinuing delightful child-rearing voluntarily required some courage. I am really grateful that I was given an opportunity to publish the book.

　I am pleased that I have done a good job supported by numerous people to the very end. The publication of this boot is a result of monthly workshops of the editorial committee of former members of the Mende Seminar. Former seminar members participated in the project spending much time while they carried out their full-time work. I would acknowledge with thanks the cooperation given by Ms. Naoko Kato, chairperson of the editorial committee, Mr. Masahiro Nakamura, in charge of AD and Ms. Mayuko Tsunoda, who handled seminar data in particular. I would also be grateful to Mr. Takashi Sugimoto and Ms. Kazuko Koike, my seniors, and Mr. Nobutaka Kotake, Mr. Kenya Hara and Mr. Taku Sato, my colleagues, for their contribution to the publication. I would also like to take this opportunity to express my sincere gratitude to Mr. Koji Aikawa of Kajima Institute Publishing Co., Ltd. who provided strict but warm guidance to us despite our selfishness, and Mr. Tatsuki Takaki who cooperated in editing. Also, I added here that this publication of this book was supported by a grant from Musashino Art University.

January 6, 2013
KAORU MENDE

0期生

瀬川 佐知子
Sachiko SEGAWA
照明デザイナー

高橋 量子
Ryoko TAKAHASHI
建築系会社員

古川 愛子
Aiko FURUKAWA
照明デザイナー

1期生

天内 奈緒
Nao AMANAI

上田 夏子
Natsuko UEDA
照明デザイナー

大谷 芽依子
Meiko OHTANI
コーディネーター

小方 理奈子
Rinako OGATA

岡部 美楠子
Minako OKABE
環境衛生管理会社
役員

加藤 直子
Naoko KATO
ランドスケープ
アーキテクト

金子 隆太
Ryuta KANEKO

齋藤 敦子
Atsuko SAITO

佐渡 英泰
Hideyasu SADO
製薬会社員

高橋 桃子
Momoko TAKAHASHI
ビジュアル
マーチャンダイザー

田中 義久
Yoshihisa TANAKA
アートディレクター

田村 更紗
Sarasa TAMURA
照明プランナー

室川 藍
Ai MUROKAWA
プランナー / デザイナー

2期生

浅野 由梨
Yuri ASANO
ベビーシッター

有馬 まどか
Madoka ARIMA
インテリアショップ
スタッフ

稲垣 幸
Sachi INAGAKI
インテリア
コーディネーター

今丑 いづみ
Izumi IMAUSHI
神職

石見 学
Manabu IWAMI
ネットショップ運営

内山 統子
Noriko UCHIYAMA
クリエイティブ
ディレクター

浦埜 好美
Yoshimi URANO
高等学校美術教員

大槻 初実
Hatsumi OTSUKI

吉良 真澄
Masumi KIRA
パラリーガル

窪田 照彦
Teruhiko KUBOTA
デザイナー

齋藤 美晴
Miharu SAITO

佐藤 未麻 Miasa SATO デザイナー	鈴木 彩子 Ayako SUZUKI ネイルアーティスト	鈴木 悠平 Yuhei SUZUKI プロソシアルダンサー	竹山 泉 Izumi TAKEYAMA フォトグラファー	田中 瑞穂 Mizuho TANAKA 不動産会社員	出川 ゆり子 Yuriko DEGAWA 建築業 トータルマネージャー

3期生

濁沼 朱里 Juri NIGORINUMA ウェブデザイナー	平石 圭子 Keiko HIRAISHI 化粧品会社 デザイナー	福尾 美雪 Miyuki FUKUO フリーカメラマン	渡辺 奈緒子 Naoko WATANABE ビジュアル マーチャンダイザー		石嶋 義彦 Yoshihiko ISHIJIMA 照明デザイナー/営業
内野 春佳 Haruka UCHINO 照明デザイナー	内村 萌 Moe UCHIMURA デザイナー	河村 江莉 Eri KAWAMURA 会社員	佐野 元春 Motoharu SANO 陶芸家	澁谷 有紗 Arisa SHIBUYA 建築士	林 まりな Marina HAYASHI 専業主婦

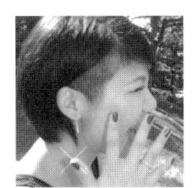

4期生

原田 裕美 Yumi HARADA 学芸員	福澤 洋樹 Hiroki FUKUZAWA 内装プランナー	矢野 大輔 Daisuke YANO 照明デザイナー	吉澤 隆介 Ryusuke YOSHIZAWA プランナー	渡辺 友理 Yuri WATANABE デザイナー	
池田 俊一 Shunichi IKEDA 照明デザイナー	長田 直子 Naoko OSADA 和菓子製造	川瀬 さや香 Sayaka KAWASE 専業主婦	久保 陽太 Yota KUBO デザイナー	小塩 俊介 Shunsuke KOSHIO ブースデザインの 営業、企画など	酒井 麻希 Maki SAKAI 照明デザイナー

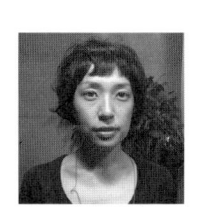

138人の面出ゼミ卒業生
138 ALUMNI OF MENDE SEMINAR

島添 拓也	蕎原 愛	寺澤 知	平本 明子	山崎 耕平	六角 望
Takuya SHIMAZOE	Ai SOHARA	Tomo TERASAWA	Akiko HIRAMOTO	Kohei YAMAZAKI	Nozomi ROKKAKU
CGデザイナー	照明プランナー	店舗設計デザイン	事務職会社員	広告代理店営業	ウェブクリエイター

5期生

青木 萌	イム ドヒ	奥田 啓晃	川島 英明	中村 将大
Moe AOKI	Dohi IMU	Hiroaki OKUDA	Hideaki KAWASHIMA	Masahiro NAKAMURA
専業主婦		デザイナー	からくり細工職人	デザイン専門学校講師

原 広大	原田 響子	平井 辰可	不殿 晴子	横井 さやか
Koudai HARA	Kyoko HARADA	Tokiyoshi HIRAI	Haruko FUDONO	Sayaka YOKOI
デザイナー	デザイナー/コピーライター	照明デザイナー	デザイナー	グラフィックデザイナー

6期生

甲斐 蓉子	川合 亮	佐々木 香枝	靏見 綾紗	中川 香里	中村 祥子
Yoko KAI	Ryo KAWAI	Kae SASAKI	Ayasa TSURUMI	Kaori NAKAGAWA	Shoko NAKAMURA
インテリアデザイナー	ウェブディレクター/デザイナー	デザイナー/アーティスト	会社員	会社員	アパレルバイヤー

7期生

前田 芳恵	山﨑 夏実	横橋 英司	浅見 美歌子	安達 みどり
Yoshie MAEDA	Natsumi YAMAZAKI	Eiji YOKOHASHI	Mikako ASAMI	Midori ADACHI
婦人靴職人見習い	ライティングプランナー	照明メーカー営業	舞台照明スタッフ	番組美術進行ディレクター

有吉 玄徳 Gentoku ARIYOSHI デザイナー	岩永 光樹 Coki IWANAGA 照明計画・設計	小出 明日菜 Asuna KOIDE 舞台照明会社員	輿石 洋平 Youhei KOSHIISHI ウェブデザイナー	小林 晴香 Haruka KOBAYASHI イラストレーター	小林 秀美 Hidemi KOBAYASHI 海外留学中
小林 真実子 Mamiko KOBAYASHI 原稿作成者	齋藤 道子 Michiko SAITO 照明デザイナー アシスタント	髙島 明子 Akiko TAKASHIMA スーツアクター	髙嶋 友唯 Yui TAKASHIMA きものアドバイザー	中原 桃子 Momoko NAKAHARA 照明器具会社員	南部 瑛美 Emi NAMBU インテリアデザイナー

8期生

堀本 あや Aya HORIMOTO ディスプレイデザイナー	本井 豊幸 Toyoyuki MOTOI 医療系大学 学生		井上 香奈子 Kanako INOUE 派遣社員	井上 祐輔 Yusuke INOUE 会社員／デザイナー	梅﨑 明花 Sayaka UMEZAKI 照明メーカー会社員

9期生

太田 芽以 Mei OHTA 建築設計事務所員	曽我部 優樹 Yuuki SOGABE 映像・グラフィック デザイナー	髙橋 梨香子 Rikako TAKAHASHI 会社員	中山 順嗣 Naotsugu NAKAYAMA 映画・TV業界 美術部	吉藤 かな子 Kanako YOSHIFUJI 新聞社 編集	
伊藤 一実 Hitomi ITO ディスプレイデザイナー	伊藤 史織 Shiori ITOH 美術設定デザイナー	江澤 佳世子 Kayoko EZAWA マーケティングプランナー	興松 麻美 Asami OKIMATSU カフェ製菓補助	刈谷 康時 Yasutoki KARIYA クリエイター	北野 舞 Mai KITANO ディスプレイデザイン 見習い

北村 康恵 Yasue KITAMURA 内装施工会社員	木村 光 Hikaru KIMURA 照明デザイナー	齋藤 麻里 Mari SAITO 専門学生	酒井 杏子 Kyoko SAKAI デザイナー	柴田 奈緒 Nao SHIBATA セラピスト見習い	戸井田 莉央 Rio TOIDA 内装施工会社員

10期生

廣瀬 文音 Ayane HIROSE 写真専門学生	保坂 安美 Ami HOSAKA ディスプレイデザイナー	松本 理沙 Risa MATSUMOTO 映画・TV業界 美術スタッフ	吉田 尚加 Naoka YOSHIDA 大学院生	渡邊 琢斗 Takuto WATANABE ディスプレイデザイナー	

生方 麻美 Mami UBUKATA 学生	鬼澤 涼子 Ryoko ONIZAWA 学生	永井 宏武 Hiromu NAGAI 学生	中村 悠未 Yumi NAKAMURA 学生	二反田 和樹 Kazuki NITANDA 学生	林 秀樹 Hideki HAYASHI 学生

大学院

林 倫子 Rinko HAYASHI 学生	山内 栞 Shiori YAMAUCHI 学生		車 寅虎 Ino CHA 大学院兼任教授/ 照明デザイナー	堀原 佳林 Karin HORIHARA デザイナー	古川 郁 Iku FURUKAWA アーティスト

杉浦 貴美子 Kimiko SUGIURA ライター/編集	葉 玉 Yu YE 照明デザイナー	加賀美 鋭 Satoki KAGAMI 照明デザイナー (5期在籍)	寺崎 舞 Mai TERASAKI 大学院生	キム ビッナリ Bitnary KIM 大学院生	蔡 知垠 JiEun CHAE 大学院生(9期在籍)

助手

福田 寿寛
Tochihiro FUKUDA
建築プレゼ制作 /
広報企画

下田 圭一
Keiichi SHIMODA
家具企画・設計

山下 匡紀
Masaki YAMASHITA
まちづくりディレクター

角田 真祐子
Mayuko TSUNODA
デザイナー

2002年 — 2012年男女比グラフ

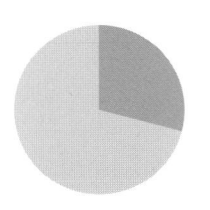

武蔵野美術大学
■ 男子　29%
■ 女子　71%

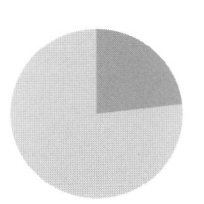

空間演出デザイン学科
■ 男子　23%
■ 女子　77%

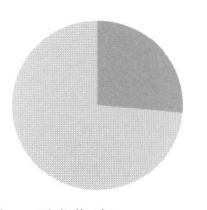

面出薫ゼミ
■ 男子　29%
■ 女子　71%

面出薫ゼミ血液型比 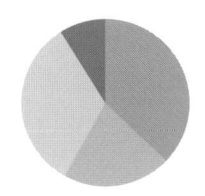
■ A　37%
■ B　21%
■ O　33%
■ AB　9%

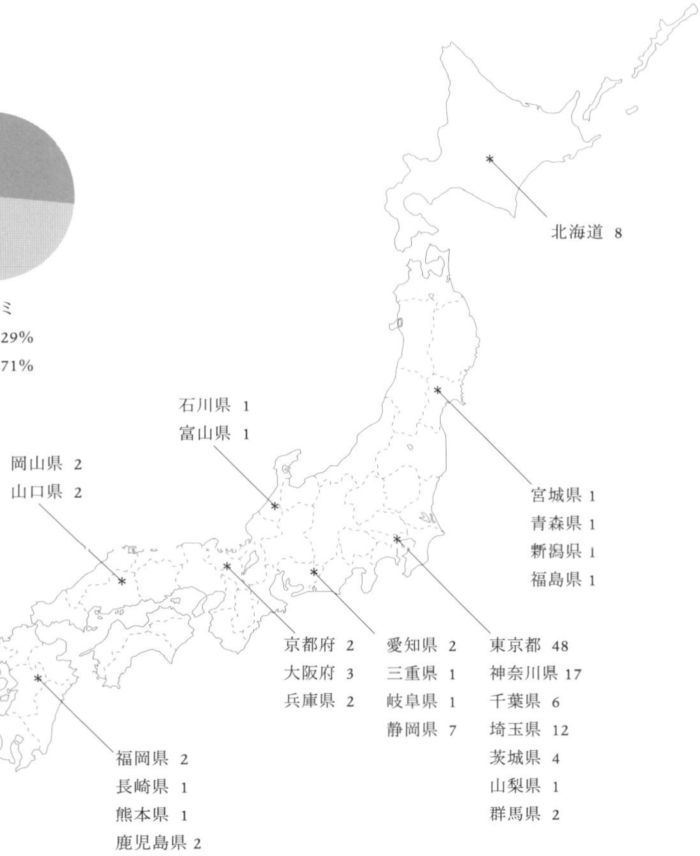

石川県 1
富山県 1

岡山県 2
山口県 2

宮城県 1
青森県 1
新潟県 1
福島県 1

北海道 8

京都府 2
大阪府 3
兵庫県 2

愛知県 2
三重県 1
岐阜県 1
静岡県 7

東京都 48
神奈川県 17
千葉県 6
埼玉県 12
茨城県 4
山梨県 1
群馬県 2

福岡県 2
長崎県 1
熊本県 1
鹿児島県 2

中国 1
韓国 4

単位（人）

138人の面出ゼミ卒業生
138 ALUMNI OF MENDE SEMINAR

243

光のゼミナール編集委員会
Profiles of editorial committee members

面出薫　*Kaoru Mende*

1950年東京に生まれる。東京芸術大学大学院美術研究科修士課程修了。1990年、株式会社ライティングプランナーズ アソシエーツ（LPA）を設立、代表取締役。主に建築照明、都市・環境照明デザインの分野で実績を持ち、市民参加の実践的照明文化研究会「照明探偵団」を主宰する。東京国際フォーラム、JR京都駅ビル、六本木ヒルズ、シンガポール中心市街地照明マスタープラン、JR東京駅丸の内復原駅舎などの照明計画を担当。国際照明デザイン最優秀賞、日本文化デザイン賞、毎日デザイン賞などを受賞。著書に『世界照明探偵団』鹿島出版会、『陰影のデザイン』六耀社など。2002年4月より2013年3月まで武蔵野美術大学教授として教鞭をとる。

Kaoru Mende was born in Tokyo in 1950, earned a bachelors and masters degree from Tokyo University of Art. In 1990, he founded Lighting Planners Associates Inc (LPA). His design ranges widely from architectural to urban environmental lighting. He is also the acting chief of the "Lighting Detectives", a citizens' group that specializes in the study of the culture of lighting, and has been involved in such superb projects as Tokyo International Forum, JR Kyoto Station, Roppongi Hills, Singapore City Center Lighting Master Plan and Tokyo station. His numerous awards include International Association of Lighting Designers (IALD) Radiance Award, Japan Culture Design Award, Mainichi Design Award. Books he has authored include, "Transnational Lighting Detectives" (Kajima Publishing), "Designing with Shadow"(Rikuyosya). Mende had been a professor at Musashino Art University since April 2002 till March 2013.

編集委員長
Chair of Editorial committee

加藤直子 Naoko Kato

1981年京都市生まれ。ランドスケープ・アーキテクト。2004年に武蔵野美術大学を卒業後、文部科学省の奨学生として米国に留学。ペンシルバニア大学デザインスクール修士課程を2009年に修了。現在、(株)フィールドフォー・デザインオフィスにて勤務。コイズミ国際学生照明デザインコンペ金賞受賞ほか。日本造園学会会員国際委員。

Born in Kyoto in 1981. Landscape Architect. Graduated from Musashino Art University in 2004 and studied in the United States as a Ministry of Education, Culture, Sports, Science and Technology of Japan scholarship student. Finished her master's degree at school of Design, University of Pennsylvania in 2009. Currently works for Field Four Design Office. Won a gold prize in the Koizumi International Lighting Design Competition for Students and other awards. Member of the International Committee of the Japanese Institute of Landscape Architecture.

中村将大 Masahiro Nakamura

1983年福岡県大牟田市生まれ。デザイン専門学校職員。2008年 武蔵野美術大学卒業。2009年 学校法人専門学校 東洋美術学校デザイン研究室 着任。2011年 同 産学連携事務局。2009年 朗文堂タイポグラフィ・スクール新宿私塾 修了。

Born in the city of Ohmuta, Fukuoka Prefecture in 1983. Member of a design academy. Graduated from Musashino Art University in 2008. Obtained a post at the Design Laboratory of the Toyo Institute of Art and Design in 2009 and transferred to the Industry-University Cooperation Secretariat in 2011. Completed courses in Robundo Typography School in "Shinjuku-Shijuku" in 2009.

角田真祐子 Mayuko Tsunoda

1985年神奈川県三浦市生まれ。武蔵野美術大学空間演出デザイン学科研究室助手・デザイナー。2008年武蔵野美術大学を卒業。2008年武蔵野美術大学空間演出デザイン学科研究室教務補助員着任。2009年同研究室助手就任。2009年〜2013年まで面出薫ゼミを担当。2009年デザインユニットminna設立。GOOD DESIGN賞 受賞ほか。

Born in the city of Miura, Kanagawa Prefecture in 1985. Assistant Professor of the Department of Scenography, Display & Fashion Design, Musashino Art University. Designer. Graduated from Musashino Art University in 2008. Was assigned a job of supporting instructor of the Department of Scenography, Display & Fashion Design in 2008 and was appointed an assistant professor in 2009. Was in charge of the Kaoru Mende Seminar in 2009 through 2013. Founded minna, a design unit, in 2009. Received GOOD DESIGN and other awards.

池田俊一 Shunichi Ikeda

1983年千葉県生まれ。照明デザイナー。2007年に武蔵野美術大学を卒業。照明器具メーカーのデザイン室を経て、現在、株式会社ライティングプランナーズ アソシエーツ勤務。国際照明デザイナー協会(IALD) 会員。

Born in Chiba Prefecture in 1983. Lighting designer. Graduated from Musashino Art University in 2007. Worked in a design laboratory of a lighting equipment maker and now works for Lighting Planners Associates. Member of the International Association of Lighting Designers (IALD).

不殿晴子 Haruko Fudono

1985年東京都生まれ。デザイナー。BloomWorks株式会社勤務。2008年武蔵野美術大学卒業後、北山創造研究所入社。代表 北山孝雄のアシスタントおよび、企画、デザインなどを手がける。2012年11月より現職。

Born in Tokyo in 1985. Designer. Works for Bloom Works Inc. Joined Kitayama & Company after graduating from Musashino Art University in 2008. Worked as an assistant to Takao Kitayama, company representative, and engaged in planning and design. Obtained the present post in November 2012.

六角望 Nozomi Rokkaku

1984年東京都生まれ。ウェブクリエイター。2007年に武蔵野美術大学を卒業。2007年度・2008年度・2011年度と、同大学空間演出デザイン学科研究室教務補助員として学科ウェブサイトの管理・運営などを担当。現在、フリーランスとして活動中。

Born in Tokyo in 1984. Web creator. Graduated from Musashino Art University in 2007. Administered and managed the website of the Department of Scenography, Display & Fashion Design of the university as a supporting instructor in academic years 2007, 2008 and 2011. Currently works on freelance basis.

山下匡紀 Masaki Yamashita

1979年東京都杉並区生まれ。まちづくりディレクター。ヘルシンキ美術デザイン大学への交換留学を経て2004年武蔵野美術大学を卒業。2005〜2009年同大学空間演出デザイン学科研究室助手担当。現在(株)地域交流センター企画取締役、NPO法人地域交流センター研究員。

Born in Suginami Ward, Tokyo in 1979. Community planning director. Studied at the University of Art and Design Helsinki as an exchange student and graduated from Musashino Art University in 2004. Assistant Professor of the Department of Scenography, Display & Fashion Design in 2005 through 2009. At present, Director, Regional Exchange Center Planning and researcher, NPO Regional Exchange Center.

下田圭一 Keiichi Shimoda

1978年広島県呉市生まれ。家具デザイナー。2002年に武蔵野美術大学を卒業。2003年 武蔵野美術大学造形学部空間演出デザイン学科研究室助手 着任。2007年 退任。現在、株式会社コンプレックス・ユニバーサル・ファニチャー・サプライ勤務。

Born in the city of Kure, Hiroshima Prefecture in 1978. Furniture designer. Graduated from Musashino Art University in 2002. Assistant Professor of the Department of Scenography, Display & Fashion Design in 2003. Quit the Department in 2007. Currently works for Complex Universal Furniture Supply Inc.

編集委員あとがき　*Editor's note*

2年前の冬、面出先生からの突然のメールに驚いた。そこには2013年3月をもって武蔵美を退任する知らせと、10年間を振り返る『光のゼミナール』の企画について書かれていた。

意気込んで本づくりを始めたものの、10年間分の記録写真の膨大さに唖然とした。日本に限らず世界各地に飛びまわったゼミは、多くの収穫物を携えていた。本づくりはその収穫物をもう一度味わい、光のデザインとは何だったのかを語り合うプロセスだった。

ただ、プロセスの中で、ゼミ1期生の私には経験のない課題がほとんどであることが分かり消沈する。先生が赴任して間もない初期のゼミ生は、その後の看板授業となる課題をほとんど経験していない。10年間の系譜を追いながら、面出ゼミという学び舎の苦悩と発展を見ることができたが、後輩たちがうらやましいのはいうまでもない。

卒業後も私は面出先生の背中を追いながらずっと走り続けている。10年の時を経て出版の機会を与えてくださった先生には厚く感謝申し上げたい。掲載資料にはOBのみなさんから多くの協力をいただいた。ここに御礼申し上げたい。

　　　　　　　　　　　　　　　　　　加藤直子

I suddenly received e-mail from Professor Mende in winter two years ago. It notified the planned resignation from Musashino Art University in March 2013 and described a plan to publish "Light seminar" that would recollect the ten-year period he spent in the university. I enthusiastically started producing the book. I was dumbfounded at the quantity of record photographs collected in the ten-year period. The seminar members traveled not only in Japan but throughout the world and obtained numerous fruits. The production of the book was a process of re-tasting the fruits and discussing what light design was.

I, however, was depressed in the process as I found that most of the assignments were unfamiliar to me, the first-year member of the seminar. The seminar members in the early years in the history of the Mende Seminar had experienced few assignments that later became trademark courses of the seminar. I could go through the agony and development of the Mende Seminar, a learning experience, tracing the ten-year records. Needles to say, I envy my juniors.

I have been running after Professor Mende since I graduated from university. I am deeply grateful to Professor Mende for proving me with an opportunity to publish a book after ten years' time. Numerous former members of the seminar kindly provided materials for publication. I am also thankful for them.

Naoko Kato

協力一覧:

英訳	渡辺 洋, (株)テトラ, 中山レイチェル
編集担当	加藤直子, 中村将大
資料担当	角田真祐子, 東 悟子
写真提供	金子俊男, 青木信之, 佐治康生

編集協力	甲斐蓉子	渡辺琢斗
	中村祥子	二反田和樹
	刈谷康時	永井宏武
	木村 光	加久本真美

光のゼミナール
THE LIGHT SEMINAR

武蔵野美術大学 空間演出デザイン学科 面出ゼミ10年間の記録
Ten years activities of Kaoru Mende Seminar
Department of Space Design, Musashino Art University

発行　2013年3月30日 第1刷

編著者　面出 薫＋ゼミ編集委員会
発行者　鹿島光一
発行所　鹿島出版会
　　　　〒104-0028　東京都中央区八重洲2丁目5番14号
　　　　電話 03-6202-5200　振替 00160-2-180883
書籍形成　中村将大
DTPオペレーション　高木達樹
印刷・製本　壮光舎印刷

© Kaoru Mende, Mende Seminar, 2013
Printed in Japan
ISBN978-4-306-04587-3 C1070

落丁・乱丁本はお取替えいたします。
本書の無断複製(コピー)は著作権法上での例外を除き禁じられております。
また、代行業者などに依頼してスキャンやデジタル化することは、
たとえ個人や家庭内の利用を目的とする場合でも著作権法違反です。

本書の内容に関するご意見・ご感想は下記までお寄せください。
URL:http://www.kajima-publishing.co.jp
E-mail:info@kajima-publishing.co.jp